MORAL GEOGRAPHIES

to Margaret with love

MORAL GEOGRAPHIES

ETHICS IN A WORLD OF DIFFERENCE

David M. Smith

Edinburgh University Press

Edinburgh University Press Ltd
22 George Square, Edinburgh

Typeset by IMH(Cartrif), Loanhead, Scotland
Printed and bound in Great Britain by The Cromwell Press, Trowbridge, Wilts

A CIP Record for this book is available from the British Library

ISBN 0 7486 1278 5 (hardback)
ISBN 0 7486 1279 3 (paperback)

CONTENTS

PREFACE

A moral crisis threatens, or so mass media tell us. Schools are exhorted to teach the difference between right and wrong, while citizens are reminded that they have responsibilities as well as rights. This is happening against a background of loss of what were once considered moral certainties, as political and religious authority in the domain of values evaporates and as postmodern attitudes permeating academia cast doubt on any claims to truth. Contemporary Western society seems to have lost its moral bearings, except for uneasy faith in market forces, while other forms of fundamentalism appear to be consolidating elsewhere. Meanwhile, some governments pursue horrific policies like ethnic cleansing with impunity, calling into question the very idea of moral progress. And, as socialism fades into memory, the globalisation of capitalism turns out to be a mixed blessing, bestowing its benefits and burdens unevenly in a world increasingly polarised between rich and poor. If ever there was a need for public debate on moral issues, this is the time. Otherwise, humankind risks losing any residual grip on the meaning of the good life, any capacity to recognise and challenge evil.

This book is both a first and a last. It is the first integrated text to explore the interface of geography and ethics, or moral philosophy. Its immediate predecessor, a collection of papers entitled *Geography and Ethics: Journeys in a Moral Terrain* (1999), co-edited with James Proctor, conveys something of the contemporary geographical engagement with ethics. The present work seeks to consolidate this new direction, taken by (predominantly) human geographers recognising common ground with a relatively unexplored field of inquiry. Not before time for a discipline which has well-established links with an array of social sciences, including economics, sociology and politics, and with the multi-disciplinary fields of cultural, development, environmental, regional and urban studies. And not before time for a discipline the practitioners of which are far from reticent in making value judgements and claims to social relevance.

This is also the last in a sequence of books in which I have set out perspectives on human geography with a normative emphasis. Initially, the framework was that of the social indicators movement, concerned with measuring the quality of life and its improvement, in *The Geography of Social Well-being in the United States* (1973). This approach was subsequently refined by the conceptual apparatus of welfare economics, in *Human Geography: A Welfare Approach* (1977). Then came a more empirical turn, towards spatial inequality or uneven development, in *Geography, Inequality and Society* (1987), elaborated in specific contexts in *Apartheid in South Africa* (1985) and *Urban Inequality under Socialism* (1989). All this culminated in the exploration of alternative approaches to social justice and their geographical significance, in *Geography and Social Justice* (1994). A growing awareness of the interdependence of questions of justice and of the good life, and of the deeply geographical nature of moral issues, led to the present volume.

The structure within which I work recognises the abiding significance of concepts central to geographical inquiry: landscape, location and place, locality, proximity and distance, space and territory, development and nature. Their elaboration has been driven to some extent by my own interests, and by the case material assembled as the project was forming. If the individuality of creative scholarship has taken me in some different directions from those which others might have followed, this is part of the diversity that continues to make geography so exciting.

The importance of the subject and its contemporary relevance have strengthened my usual inclination to write with students in mind. My aim has thus been an accessible text that could be used in advanced undergraduate or postgraduate courses with an orientation towards the ethical and moral. In trying to engage complex issues without undue simplification, I also have regard to professional peers in geography and cognate subjects. Moral philosophers may find points of interest, though they will soon see that I am not one of them. At the very least, this book should help to explain how geographical context is significant to moral practice, and how ethical deliberation is incomplete without recognition of the geographical dimension of human existence.

ACKNOWLEDGEMENTS

Some of the research on which this book is based goes back many years. Various projects have benefited from the support of the British Academy (thanks especially to Jane Lyddon), the British Council and the Economic and Social Research Council. Completion of fieldwork and library research, and substantial progress on the text itself, was greatly facilitated by a Leverhulme Fellowship, allowing the luxury of some travel and research assistance which would not otherwise have been possible.

A year's sabbatical leave granted by Queen Mary and Westfield College was also helpful, as were small grants from the Social Science Faculty Research Fund. Successive heads of the Department of Geography have let me get on with my own thing, for the most part, while other colleagues have tolerated my single-minded commitment to this project. I am grateful to them all.

I owe special thanks to James Proctor, and to others who participated in the trans-Atlantic Geography/Ethics Project which generated our edited collection *Geography and Ethics*. Working with Jim and our contributors helped to broaden my knowledge, and Jim kindly reviewed Chapter 9 in draft. I am also grateful to members of the Critical Geography Forum and others who responded to requests for references and reprints. An invitation from Rex Honey to lecture at the University of Iowa, at a crucial time, strengthened my resolve.

Some parts of the book are based on papers already published in journals. I am grateful for the editorial encouragement of Peter Dicken, Ron Johnston, Dick Peet, John Silk and Tim Unwin, and to reviewers for their constructive critiques of the original manuscripts, as well as to the publishers for permission to draw on these works. The contribution of this material to relevant chapters is recognised in the text, but I also acknowledge these papers here: 'Geography and moral philosophy: some common ground' and 'Social justice and the ethics of development in post-apartheid South Africa' in *Ethics, Place and Environment*; 'How far should we care? On the spatial scope of beneficence', 'Moral progress in human geography: transcending the place of good fortune', and two progress reports on geography and ethics in *Progress in Human Geography*; and 'Geography, community, and morality' and 'Social justice revisited' in *Environment and Planning A*.

As a matter of style, I have used quotations more than is my customary practice. This is because it seems right for crucial ideas to be expressed in the carefully chosen words of the writers concerned. The sources of all quotations are cited, but I take this opportunity to acknowledge a particular debt to writers on whom I have drawn most freely, and from whom I have learned more than I am able to show.

Numerous colleagues in different parts of the world have provided various kinds of help in the development of my case material. In the process, they have continually renewed my faith that the mutuality of scholarly cooperation will survive the parochial self-interest encouraged by the contemporary fetish of performance assessment. This has been an ethic of care in practice, transcending divisions of culture and politics as well as distance.

In Poland, these colleagues include Grzegorz Węcławowicz, who introduced me to the city of Łódź (as well as hosting visits to Warsaw), Professor Stanisław Liczewski, who made me welcome at the University of Łódź, and Sylwia Kaczmarek who shared her knowledge of the city's industrial history and architecture. Marek Jerczińsky, Piotr Korcelli and Alina Muzioł-Węcławowicz are among others to whom I have become indebted, over the years, for access to their research and hospitality. I am particularly grateful to Bolek Domański of the Jagiellonian University, Kraków, from whose work on the industrial city I have learned much, and whose introduction to the former Kazimierz ghetto helped to stimulate my interest in Polish-Jewish relations. And to the memory of Maikol Rajna from Wrocław, family friend, one of many displaced by war and never to return.

In Israel, Shlomo Hasson of the Hebrew University of Jerusalem has been a frequent guide and source of information on the city's neighbourhoods. I shall never forget the introduction to Mea Shearim by the late Shalom Reichman on a very wet day. Yosseph Shilhav helpfully kept me up to date with his research on the ultra-orthodox. Others who repeatedly made me welcome include Rafi Bar-El, Yoram Bar-Gal, Yehuda Gradus, David Grossman, Nurit Kliot, Gabi Lipshitz, Avinoam Meir, David Newman, Arie Shachar, Stanley Waterman and Oren Yiftachel. Some visits were funded by the Academic Study Group on Israel and the Middle East (thanks to John Levy). Special acknowledgement is due to Amiram Gonen, director of the Floersheimer Institute for Policy Studies in Jerusalem, whose reaction to my first attempt to write about the Łódź ghetto encouraged me in further explorations which led to the weaving of Jewish experience through four chapters of the book.

In South Africa, Dhiru Soni, Brij Maharaj, Vadi Moodley and their colleagues in the Geography Department at the University of Durban-Westville made me welcome on regular visits in recent years. Earlier, Ron Davies at the University of Cape Town, Kieth Beavon at the University of the Witwatersrand, and Roseanne Diab and Gerry Garrard at the University of Natal, Durban, provided similar support. Numerous others whose help has been significant include Rob Fincham, Linda Grant, Skip Krige and Jeff McCarthy. It has also been instructive to learn from the work of a new generation of South African geographers, including Meshack Khosa, Udesh Pillay and Jenny Robinson. Above all, I am grateful to Denis Fair, for more than three decades of intellectual stimulation and friendship.

Having acknowledged all this assistance, accepting personal responsibility for the interpretations offered in this book goes much further than the conventional disclaimer. I have dealt with controversial and sensitive matters, as an outsider, and I would not be surprised if some of those who helped would take issue with me.

In assimilating the results of some years of field and library work, and in building up to the text, research assistance provided by Priscilla Cunnan, Rob Higham and Tracey Smith relieved me of substantial effort finding references, trawling the Internet, and undertaking other tasks.

This book might not have been published but for the encouragement of John Davey, my editor of choice for more than twenty years, whose support, through this Press, was crucial to my own committment to see the project through.

Finally, to my nearest and dearest, for their active assistance, including well-chosen gifts which have extended my reading. Tracey shared some of the fieldwork (most memorably in Jerusalem), and helped all the way through to proof reading. Michael provided regular respite care at Soho House, as well as his customary backup to a technologically challenged Mac-user. Margaret's love saw me through dark days; her response to the awesome grandeur of the industrial landscape of Łódź brought full circle a journey we began among the mills of the Derbyshire Dales more than forty years ago, and which I would not have completed without her.

David M. Smith
Loughton, Essex, 16 July 1999

At each time and place, those who hope to move towards justice and towards virtue will have to build, and to rebuild, shaping the institutions, polities and practices which they find around them, and their own attitudes and activities, to meet standards which, they believe, can be standards for all within the domain of their ethical consideration. (Onora O'Neill [1996], *Towards Justice and Virtue: A Constructive Account of Practical Reasoning*: 212)

As far as the work of ethical theory and practice is concerned, useful and accurate notions of human flourishing can only emerge from richly contextualized, sometimes local, evaluations of what it means to be human, what people want and strive for, and what enables their living in ways they value in specific historical and cultural locations. (Chris J. Cuomo [1998], *Feminism and Ecological Communities: An Ethic of Flourishing*: 79)

CHAPTER 1

INTRODUCTION: GEOGRAPHY, MORALITY AND ETHICS

> Human beings do not impose alterations on nature merely to survive. They aspire beyond mere survival to the good – good human relations and a good place in which to live. (Tuan 1989: viii)

Human beings have moral values. That is to say, we have ideas about what is right or wrong, good or bad, better or worse, in connection with important aspects of life. These values guide our actions, helping us to decide what we should or should not do, how we ought to live; and they provide a basis for evaluation of the conduct of others. They can vary among individuals and groups, and hence from place to place, as part of our world of difference. Furthermore, these values are not fixed or immutable: we can reflect on them, by ourselves and in discourse with other persons, and we may change them. It is this capacity to reason about normative aspects of life, beyond the pursuit of mere physical survival, that most clearly differentiates humankind from other creatures. This is how we are, how we have become. It is part of our distinctively human nature.

MORAL VALUES IN GEOGRAPHY

It is sometimes supposed that science is concerned with discovering what is, what actually exists, rather than what ought to be. The realm of human values is left to philosophy and theology for the most part, or to those creative arts with a licence to speculate about the meaning of life. But this distinction is false. Values are implicated in science, from choice of research topics, through methods adopted, to presentation of findings. Thus, 'ethics is not so much something that should be added on to science as discovered within it' (Proctor 1998b: 295). The pursuit of space exploration, for example, like the advance of medical science, expresses the judgement that this is of some value to humankind. The activity in question may be instrumental in achieving some such good as improving telecommunications or relieving suffering. It may also have intrinsic value: finding knowledge or truth for its own sake is respected well beyond the ivory tower. The normative content of scientific method is illustrated by controversy over experiments on animals, raising the question of whether the end justifies the means. As to presentation of findings, these can be slanted in various ways, for example if they threaten the interests of sponsors or even the prejudices of scientists themselves.

In social science, including human geography, the object of inquiry is humankind, creator of the very values which permeate scientific activity and all other aspects of life. From day-to-day interaction with family, friends and neighbours, through community organisations, to institutions devised for the governance of societies, values are called into play, contested, negotiated and reassessed. They may be largely implicit, embedded in the personal understandings or community codes of morality which guide individual behaviour. They can be more explicit, as part of a political philosophy or theory of the good life. They are sometimes set down with legal precision, for example, in a national constitution, bill of rights or code of professional ethics.

In geography, which combines social and environmental science, values involve not only the behaviour of human beings towards one-another but also interaction within nature. Indeed, the very distinction between humankind and the physical environment is problematic, scientifically and also ethically: we are an integral and active part of nature, with consequent responsibilities. The utilisation of the Earth's resources is replete with normative issues, including the preservation of natural beauty, the protection of people from pollution, and responsibility to future generations. The rise of 'green' politics highlights the growing tension between the economic exploitation of nature and conservation of the environment.

The issue is not, then, whether geography is a positive or normative endeavour, but how both perspectives are inseparably implicated in trying to make sense of the world, and perhaps to improve it.

There have been times when the positive appeared to dominate the discipline. This was the case during the 'quantitative revolution' of the 1960s and the subsequent era of geography as 'spatial science'. Fashioning themselves on the prestigious physical sciences, and on the technical sophistication which earned economics so much respect, the quantifiers and model-builders tended to reduce the subtle diversity of areal differentiation to a geometry of spatial form. Their abstraction from the value-laden reality of actual human life helped to stimulate a normative reaction, in the 'radical geography' of the 1970s (D. M. Smith 1971; Peet 1977). This called for greater social relevance, at a time when the euphoria of technological progress was being deflated by discovery of the persistence of dire poverty within otherwise affluent societies as well as in the underdeveloped world. Inequality became an issue. So did social justice, as the morality of the 'development gap', political domination, social deprivation, racial discrimination and the like was increasingly called into question. Added to this was the growing perception of an environmental crisis, as pollution and depletion of resources were recognised as costs of economic growth. The quality of life became a focus for academic and political debate.

Geographical engagement with these issues brought values very much to the fore. Topics like crime, health and hunger, hitherto largely neglected, were added to the research agenda, and brought together in attempts to identify a geography of social well-being (D. M. Smith 1973; Knox 1975). This added a spatial dimension to the contemporary social indicators movement, which sought to augment measures of human progress confined to economic performance. A focus on the geographical expression of inequality prompted a spate of empirical investigations (Coates, Johnston and Knox 1975; D. M. Smith 1979). And there was an attempt to restructure the discipline around the theme of human welfare (D. M. Smith 1977).

Common targets for moral indignation were various forms of spatial inequality, including discrimination in access to facilities satisfying human needs and in proximity to sources of nuisance. It was rare, however, for the philosophical foundations of values in geography to be explored. An exception was the work of David Harvey (1973), which drew on social and moral philosophy in sketching out specifications for territorial social justice (see Chapter 7). His transition from a liberal perspective to the historical materialism of Karl Marx revealed the contested character of the values underpinning alternative ways of organising the societal process of production and distribution. Another such contribution was by Annette Buttimer (1974) who saw a humanistic approach, grounded in existential philosophy concerned with a person's mode of 'being-in-the-world', as crucial to understanding human values in their historical and social contexts.

As radical geography gathered strength during the 1970s, the influence of Marxism increased. A mechanical form of structural determination, in which cultural and social forms followed from the economic base of society, allowed little scope for the individual volition stressed by humanists. And the elaboration of territorial social justice hardly seemed necessary, so self-evident were the injustices of capitalism taken to be, and so obvious the superiority of socialism. By the time the roles of human agency and contingency had been reasserted strongly enough to penetrate Marxian metatheory, it was geography's humanist strand which appeared better prepared to deal with values.

The most sustained attention to moral issues was provided by Yi-Fu Tuan (1986, 1989, 1993). He recognised that the meaning of the good life varies greatly among cultures, as well as among individuals in complex modern societies. Yet we do share things in common: certain experiences of heightened feeling and consciousness involving relations with others are surely transcultural (Tuan 1986: 13). In considering how morality is lived and imagined in different times and places, he linked the geographer's traditional interest in transformed nature with moral-ethical systems. In so doing, he put morality at the very core of human creativity, and of geography. He argues that moral issues emerge with the human use of the Earth, which raises questions of right and wrong, good and bad in the actors as well as in those who, like geographers, observe and comment on them (Tuan 1989: vii–viii). Moral imagination calls for attention to the real, the particular, with which geography often claims to be preoccupied. But the particular cannot stand by itself; thus:

> If some people aspire to the true and the good by way of the concrete particular, others appear to be drawn, naturally and from the start, to the true and the good as embodied in abstract theories and principles. The opposite of the concrete particular is not necessarily swooning fantasy or the reductive image of an egotistic mind; it can be something 'hard' (in the sense of necessary and compelling) that combines the greatest generality with precision. (Tuan 1989: 178)

Landscape interpretation has been an abiding disciplinary interest (see Chapter 3), to which Tuan has given a distinctive moral twist. He suggests that, when we are preoccupied with our senses, it is easy to forget cultural institutions and social arrangements, and the material base that affects them. But when we shift attention to such artefacts as paintings, sculptures, landscape gardens, buildings and cities it is no longer so easy to repress moral questions. He asked whether magnificent buildings and ceremonies are a sign of injustice in the social realm, whether 'conspicuous public art, by its commanding presence, has served historically to distract people from comprehending their own oppressed condition' (Tuan 1993: 214–15). He sees rivalry between the good and the beautiful, and concludes that culture 'is a moral-aesthetic venture, to be judged ultimately by its moral beauty' (Tuan 1993: 240).

These references to the work of Yi-Fu Tuan raise three crucial issues, or tensions, at the interface of geography and ethics. One is between evident cultural differences in conceptions of the good and some apparently transcultural moral experiences. Another is between the particular and the abstract as sources of ethical understanding. The third is between the moral and the aesthetic, as expressed in the human creation of culture and especially of landscape. These will all be revisited, in various ways, as this text unfolds.

By the beginning of the 1990s the concern with values in geography, inherited from the era of radicalism, was being augmented from a number of directions. This prompted sessions at conferences and special issues of journals (*Urban Geography,* 15, 7 [1994]; *Antipode,* 28, 2 [1996]; *Society and Space,* 15, 1 [1997]). Other indications of this 'moral turn' include the first volume in a geography and philosophy series – on environmental ethics (Light and Smith 1997), a new journal *Ethics, Place and Environment,* an edited collection on topics connecting geography with ethics (Proctor and Smith 1999) and reviews of the growing literature at this disciplinary interface (Proctor 1998a; D. M. Smith 1997b, 1998a, 1999b). There is even a suggestion that geography might have a role in moral education (D. M. Smith 1995d).

One of the most influential directions of moral inquiry came from a reinvigorated cultural and social geography. This extended an earlier preoccupation with race(ism) to concerns about the disadvantage of various 'others', subject to unfair treatment by virtue of societal failure adequately to acknowledge and to respond to their differences. These groups include women, cultural or ethnic minorities, the disabled, those whose sexual orientation departs from the perceived norm, and postcolonial subjects. The emphasis on difference reflects one of the prime concerns of postmodernism, with its hostility to the essentialising (homogenising and universalising) tendencies inherited from the Enlightenment.

The application of a 'moral lens' to human geography was anticipated as follows by the Social and Cultural Geography Study Group of the Institute of British Geographers, in an argument for (re)connecting their inquiries to moral philosophy:

> [W]e will seek to establish the geography *of* everyday moralities given by the different moral assumptions and supporting arguments that particular peoples in particular places make about 'good' and 'bad'/'right' and 'wrong'/'just' and 'unjust'/'worthy' and 'unworthy'. There can be little doubt that these assumptions and arguments *do* vary considerably from one nation to the next, from one community to the next, from one street to the next. (Philo 1991: 16)

These variations overlap with differences of social class, ethnic status, religious belief, political affiliation and so on: moral assumptions are bound up with the social construction of such groups, with who is included and excluded. Attention was also drawn to the geography within everyday moralities: 'moral assumptions and arguments often have built into their very heart thinking about space, place, environment, landscape'. Local culture is closely implicated in the geography of everyday moralities, 'which "glue" together the assumptions and arguments of particular peoples in particular places' (Philo 1991: 19). And culture is a site of contest and conflict, with moralities made and remade across space.

Thus, the relationship between local moral beliefs and practices, society and culture was problematised, in various ways. Particularly important was the recognition that 'recent debates – over Marxism, humanism and postmodernism, for example – have radically transformed the context of deliberations over the place of morality in geography' (Driver 1991: 61). Power, representation and resistance have become keywords in the analysis of how some persons or groups seek to have their way over others.

Work conducted under the rubric of 'moral geographies' has subsequently become common enough to be featured in reviews of both social and cultural geography (Matless 1995: 396–7) and political geography (Ó Tuathail 1996: 409–10). These studies provide a variety of moral readings of human behaviour in relation to the built environment and also to nature (see Chapter 3).

Another direction from which moral issues have (re)surfaced is the return of social justice to the geographical agenda (D. M. Smith 1994a; see Chapter 7). Harvey (1996) has reasserted his materialist perspective in exploring 'the just production of just geographical differences'. Low and Gleeson (1998) have also written on justice at book length, sharing with Harvey the introduction of nature into a discourse hitherto confined to human aspects of the subject. The diversity of geography's new engagement with social justice is indicated in a collection of papers edited by Laws (1994), in contributions on such issues as health care (D. M. Smith 1995c), population migration (Black 1996) and urban policy (Badcock 1998), and in treatments of some theoretical problems (A. M. Hay 1995; Gleeson 1996; D. M. Smith 1997a). There is also work in the established tradition of territorial social justice (Boyne and Powell 1991, 1993).

Other directions of the moral turn will emerge during the course of this book. The one which remains to be mentioned briefly here is professional ethics. Interest in this subject has grown considerably in recent years, raising a wide range of issues (e.g. Brunn 1989; Kirkby 1991; McDowell 1994; Rose 1997; I. Hay 1998). Some have been prompted by the changing societal context of academic activity, including increasing pressure of performance assessment and commodification of knowledge. The importance of associating a personal or institutional name with research findings which may generate commercial value as well as professional prestige raises questions of intellectual property rights (Corry 1991). Innovations in the collection and dissemination of information, such as remote sensing, geographical information systems (GIS) and the Internet, pose ethical issues which include uneven access and intrusions on privacy (Lake 1993; Crampton 1995).

The construction of geographical knowledge is now part of the discipline's problematic. Ethical aspects of the treatment of research subjects and the representation of their lives, first raised in a sustained way almost twenty years ago (Mitchell and Draper 1982), are of particular contemporary concern. Among the issues discussed by contributors to the edited collection referred to above are the importance of a communicative ethics in participatory research, the legitimacy of persons writing about a group of which they are not members, the conduct of cross-cultural research involving encounters with alternative views of the world, and the relationship between research student and supervisor with different agendas (Proctor and Smith 1999). There is a growing recognition that research involving other persons requires a relational ethics emphasising reciprocity between researchers and researched.

The (re)discovery of moral issues by geographers may be linked to a broader normative turn in social theory (Sayer and Storper 1997). Just as the era of radical geography three decades ago reflected turbulence in the North America and Western Europe of those times of social activism and war in Vietnam, so the contemporary engagement with normative issues is associated with the moral crisis which some commentators see as enveloping societies carried away by their own (uneven) affluence and unconcerned about those excluded from the benefits. And unease is by no means confined to the political left: there is a powerful moralising discourse from the right, pointing to the breakdown of conventional morality manifest in rising crime, welfare dependency, calls for gay rights and the multiplication of households outside marriage.

The normative turn is evident in economics. Amartya Sen (1987: 7, 89) has claimed that modern economics has been impoverished by its distance from ethics, and argues for closer contact between the two fields, exemplified by his own work on poverty which earned a Nobel Prize in 1998. He also argues that narrow assumptions of self-interest have limited the scope of predictive economics, making it difficult to incorporate behavioural versatility manifest in bonhomie and sympathy for others (Sen 1987: 79). In similar vein, attention has been drawn to the role of gifts within an 'economy of regard' (Offer 1997). Others have proposed a new economics focused on the moral dimension (Etzioni 1988) and noted a revival of interest among economists in distribution (Le Grand 1991).

Interest in ethical issues has also been building up in urban and regional planning (e.g. Beatley 1994; Howe 1994; Hendler 1995). Hillier (1998) links the representation of nature in land use planning to environmental ethics, while others have taken up environmental justice (see Chapter 9). The normativity of urban and regional planning, concerning the 'right' use of land, involves a disciplinary interface in which geography has long-standing interests. The moral dimension of development, or development ethics, has attracted much attention in recent years (see Chapter 8). There is also a connection with ethical dimensions of international relations (e.g. P. J. Anderson 1996; Graham 1997), and justice among nations (e.g. Attfield and Wilkins 1992; Thompson 1992).

Another discipline of geographical interest (re)discovering ethics is anthropology. In her introduction to a collection of papers, Signe Howell (1997: 8) claims that few anthropologists have attempted the empirical study of moral discourse, and its theoretical challenges, adding that philosophers are not concerned with 'locating the moral subject within social and cultural worlds'. Some of the ethnographic approaches to morality in her case studies might provide models for geographical research (see Chapter 2).

The growing interest in the interface between geography and ethics has stimulated contributions linked to interdisciplinary concerns. For example, a collection of papers on normative issues (Sayer and Storper 1997) includes contributions stressing the importance of situating discussions of justice, equality and power within an international geopolitical context (Slater 1997) and examining arguments for extending ethics to subjects beyond the human (Whatmore 1997). Research on the 'moral economic geographies' formed by participants in local exchange employment and trading systems (LETS) shows how new forms of economic and social relations mediated by 'virtual currencies' might challenge the customary preoccupations of economics (Lee 1996). Thus, geographers are beginning to participate in a more general, multidisciplinary search for ways of engaging with ethics.

The significance of the moral turn was confirmed in a presidential address to the Association of American Geographers, in which Stephen Birdsall (1996: 619, 620) concerned himself with 'everyday experiences whose characteristics reflect our patterns of regard, respect, and consequent responsibility'. He suggests: 'Our choices are, in effect, guided by a map of moral alternatives, a map of which we are not aware. Through our everyday interactions, we trace the moral geography of our lives'. His sense of moral geography as 'patterns resulting from everyday expressions of good and evil' (Birdsall 1996: 627, note 3) adds to the meanings assigned to this concept in research practice.

The most sustained recent consideration of the moral content of human geography is in the framework for understanding *Homo Geographicus* elaborated by Robert Sack. He argues that geography is at the foundation of moral judgement: 'Thinking geographically heightens our moral concerns; it makes clear that moral goals must be set and justified by us in places and as inhabitants of a world' (Sack 1997: 24). He sees the moral force of a place as its capacity to tie together the particular virtues or moral concerns of truth, justice and the natural, which exist in different and changing mixes in different places. The most graphic expression of the moral significance of place is the image of 'thick' and 'thin'. As boundaries become porous, this thins out the meaning of place, and the virtues therein, changing the thicker places of premodern society with their strongly partial moral codes. Thus:

> [T]he local and contextual should be thin and porous enough not to interfere with our ability to attain an expanded view, and the local can be understood and accorded respect only if people attain a more objective perspective, enabling them to see beyond their own partiality and to be held responsible for this larger domain . . . thick places create differences, but when they are too thick and their boundaries impermeable they prevent us from transcending them and seeing clearly.
>
> (Sack 1997: 248, 257)

Sack depicts a moral perspective graphically, operating along an axis representing movement from a local, partial perspective to one which is more global and impartial. Transcending partiality is part of growing up, of expanding horizons, of knowing more about the world and its peoples and the consequences of our actions: 'A moral position must be justified to others on the basis of a less partial or impartial reason, not on self-interest, custom, or practice' (Sack 1997: 6). Awareness of the implications of our actions within a context wider than the local provides a basis for their evaluation, and a source of moral motivation. Thus: 'we are geographical and will improve as we think geographically' (Sack 1997: 252). Implicit in this is that we will improve ourselves, as well as the lives of others.

Sack's book underlines some important issues at the interface of geography and moral inquiry, of the kind suggested above by Yi-Fu Tuan. These focus on the distinction between the local or particular manifest in specific places, and the universal or general towards which human understanding is often drawn. They point to a fundamental and deeply geographical distinction in morality: between sympathy for close and familar persons and concern for distant and different others. This distinction is sometimes construed as between a morality of feeling manifest within the confines of private space and a morality of reason expressed across the more extensive public space. But before such notions are considered further, a brief excursion into moral philosophy is required.

ETHICS, MORALITY AND MORAL PHILOSOPHY

So far, the terms values, ethics and morality have been used in an unreflective way. It is now time to be more specific, about what they mean, and about the nature of moral enquiry. What follows is inevitably brief and simplified, but sufficient to introduce what geography is engaging with at the interface of moral philosophy.

Values refer to some desirable or worthy aspect of life, something to which value is assigned. A distinction is often made between the moral values of good and right, and the non-moral values of beauty and truth, which might suggest the independence of a moral dimension of life from aesthetic and scientific dimensions. However, there are interconnections, for example, in the moral content of science and in links between aesthetic and moral readings of landscape referred to above. Another common distinction is between moral goods or virtues like courage, duty and justice, the performance of which is morally admirable, and non-moral goods like freedom, happiness and security, which are desirable experiences but not sources of moral credit. For something to be valued by human beings as moral agents, it must be human, necessary for human life, capable of making human life better, or something that humans can appreciate or respect; morally valuable entities must be capable of having interests or doing well, and include whatever is capable 'of being harmed, exploited, oppressed, degraded, pained, and mistreated' (Cuomo 1998: 48).

The terms ethics and morality are often interchangeable in everyday usage. One may be preferred to the other in particular contexts; for example, it is customary to refer to sexual morality but to medical ethics, yet both apply to codes of conduct in specific spheres of life. Two meanings of ethics may be recognised within contemporary geography: professional ethics and the broader usage of the term as moral philosophy (Proctor 1998a). The distinction to be adopted in this book, except insofar as work discussed departs from it, is between ethics as moral theory and morality as practical action (Rauche 1985: 252–3). Thus, ethics is the same as moral philosophy, or 'the conscious reflection on our moral beliefs' (Hinman 1994: 5). Morality is what people actually believe and do, or the rules they follow.

What is at stake in ethics and morality can be further elaborated as follows:

> A moral or ethical statement may assert that some particular action is right or wrong; or that actions of certain kinds are so; it may offer a distinction between good and bad characters or dispositions; or it may propound some broad principle from which many more detailed judgements of these sorts might be inferred – for example, that we ought always to aim at the greatest general happiness, or try to minimize the total suffering of all sentient beings, or devote ourselves wholly to the service of God, or that it is right and proper for everyone to look after himself [*sic*]. (Mackie 1977: 9)

The distinction between actions and characters is between 'doing' and 'being': doing the right things and being a good person. A different distinction traced back to Aristotle is sometimes made: between doing right or leading a good life and faring well or having a good life (Graham 1990: 92–5). This might distinguish between an other-regarding and a self-regarding disposition. Some considered integration of 'leading' and 'having' might point to the good in a person's whole life.

Of course, the evaluative terms of good and bad, right and wrong, can be used to describe behaviour in many aspects of life, some of them quite trivial. They may refer to what is merely prudent, such as arriving at the station in time to catch a train; not to do so can make life difficult, but would hardly be considered immoral. They may refer to obedience to prevailing law, which is the prudent thing to do so as to avoid punishment, but leaves open the question of whether the law can be defended from a moral point of view. For example, the laws prohibiting black people from using 'public' facilities under apartheid in South Africa or in the era of racial segregation in the United States were widely regarded as unjust, and those who challenged them gained moral credit. Some actions are deemed good or bad as a matter of manners, or being polite to others; failure to conform to conventions like shaking hands or thanking hosts could be thought rude, but not usually immoral.

So, when actions or dispositions are described in terms of morality, they take on special significance. Moral rules are seen as more authoritative and objective than government laws or rules of etiquette, although their source of authority and the possibility of their objectivity are matters of contention (Mackie 1977: 17). The force of morality is captured by the common proposition that it is overriding with respect to all other considerations, 'not just one more point of view but, rather, something to which all points of view must answer', and pervasive in the sense that moral considerations have 'the ability to permeate or diffuse themselves into nearly all arenas of human life' (Louden 1992: 62, 63). Thus, 'moral attitudes are more than (merely) attitudes . . . they are attitudes which we get disturbed about . . . it matters to us to secure similarity of attitudes within society' (Williams 1972: 31).

This last suggestion has implications for the spatial scope of morality. Even if we accept that morality 'is not to be discovered but to be made' (Mackie 1977: 106), that moral attitudes are social in origin if not altogether culturally specific, their authority gains strength with the extent of their reach. The difference between law and justice lies in their scope:

> [T]he comprehensive authority claimed by morals is much more extensive than that claimed by the law. The latter's authority is limited to all persons within a realm, usually territorial, and it is normally restricted to their behavior. Morality, by contrast, is imperial, claiming jurisdiction over all agents, regardless of place of

> residence, and governing intentions, feelings, and dispositions in addition to actions.
> (Morris 1996: 832)

At the extreme, the strength of morality may be invoked as natural law: an unchanging normative order that is part of the natural world and hence universal, as in Aristotle's distinction between legal or conventional justice and natural justice which has the same force everywhere (Buckle 1993: 162). The recognition of moral pluralism, and of relativism (see next section), dilutes the power of a morality which otherwise claims wide if not necessarily universal scope.

Before moving on to broad principles (as referred to in the quotation from Mackie above), we should note a conventional division within moral inquiry. This is between descriptive ethics, normative ethics and metaethics. Descriptive ethics is concerned with the identification of actual moral practices and beliefs (including ethical understandings), which can involve differences and similarities among cultures, times and places. Normative ethics is concerned with solutions to moral problems, which can include those associated with social justice, development and the environment, for example. A distinction is sometimes made between two levels of normative ethics: normative theory which looks for very general moral principles about how we ought to live and applied normative ethics which is concerned with specific issues like abortion and euthanasia. Metaethics considers what it means to engage in ethics: what moral argument is all about.

The rest of this section introduces some issues in metaethics, which impinge on broad principles (following D. M. Smith 1997b: 584–5; 1998a: 10). Metaethics concerns questions which it would be helpful to resolve, before solutions are proposed for particular moral problems. These questions include the meaning of such terms as good and bad, right and wrong, ought or should, or the language of morals. Among other things, metaethics is concerned with distinctions between the use of the term good in a moral as opposed to a functional sense (cf. a good deed and a good tool), and of the term right in a moral as opposed to scientific sense (cf. the right deed and the right answer to a factual question). Another aspect of metaethics is 'trying to find out what would constitute explaining the fact that people hold moral views' (Harman and Thompson 1996: viii–ix), asking how they come about.

Engaging metaethics is complicated by enormous differences among philosophers themselves, such that we may wonder whether they are all talking about the same thing (M. Smith 1994: 3–4). Theoretical diversity is expresses in numerous competing 'isms'. For example, a guide to the subject (P. Singer 1991) covers the following: realism (the view that there is an objective moral reality), intuitionism (that we can know moral truth by a kind of intuition), naturalism (that moral truth can be known from sensual experience of some other property), subjectivism (that moral views are personal opinions and not objective truth), relativism (that morality is relative to a particular society or culture) and universal prescriptivism (which gives prominence to reasoning about ethical judgements). A text on ethics (Hinman 1994: 54) identifies other examples of moral theories: divine command theory (which alleges that morality is determined by God's commands), the Kantian claim that morality is a matter of having the correct intention (one that can be willed universally for all human beings), the position of the egoist (who thinks that morality is solely a matter of self-interest), utilitarianism (concerned with consequences), rights theory (which sees moral issues in terms of rights and correlative duties) and virtue theory (which maintains that morality is primarily about character and not actions). Some of these positions may have different labels in different texts: for example divine command theory may be referred to as supernaturalism. And there are other overarching perspectives, like cognitivism, which takes moral judgements to be truth claims, and objectivism, which holds that moral values have objective status as opposed to being merely what persons approve of (subjectivism) or simply proclaim as right or good (emotivism). Other positions to note are nihilism, or knowing nothing about mortality, and scepticism, which is a less radical form of doubt about moral values and which seeks moral truth (if there is any) in concrete experience rather than in the domain of theory. There are also differences over the usefulness of moral theory itself, as well as about what such theory might be like (compare Williams 1985 and Louden 1992).

Clearly, the theoretical starting point, represented by the choice among these diverse perspectives, has an important bearing on the principles which may be deployed in the resolution of any particular moral issue. This is so even with the most general distinction, between a deontological perspective (like virtue theory) which stresses duty and right conduct, and a teleological perspective (like utilitarianism), which focuses on the consequences of actions in pursuit of some ultimate good. Doing right and faring well frequently conflict. Individual and group welfare, in the sense of material living standards, may be promoted by exploiting or oppressing others, which is bad conduct. Doing one's duty may have uncomfortable or even fatal consequences, which is to fare badly, as in many a heroic tragedy.

The notion of competing and contested ways of seeking truth will be familiar to anyone aware of geography's paradigm shifts and proliferating 'isms', and of the postmodern challenge to grand theory of any kind. However, most of the large problems of metaethics, arising from the apparent incompatibility of alternative moral theories, may be set aside here, best left to philosophers. Further guidance can be found in specialised works, ranging from the cartoon format (D. Robinson and C. Garratt 1996) to the compendious (P. Singer 1991; Becker and Becker 1992), and in ethics textbooks (e.g. Graham 1990; Hinman 1994; H. J. Gensler 1998). Those problems of special geographical interest arise when ethics confronts the world of difference.

THE SIGNIFICANCE OF DIFFERENCE

It is now time to pose some questions about the significance of difference to ethics. The initial concern is not with individual differences which may be morally relevant to how persons should be treated. By difference here is meant the fact of diversity, or pluralism, in moral beliefs and practices, as they vary from place to place (and from time to time), as integral features of human cultures and ways of life. Differences in how groups of people live do not necessarily invite normative judgement. If in the North the economy is dominated by mining but in the South by farming, these are geographical facts, unlikely to arouse moral indignation. However, if the differences are between poverty and affluence, or if Northerners practice infanticide to which Southerners object, then comparative evaluation can hardly be avoided. This leads to the general question of how to compare different ways of life, involving judgements of better or worse.

There is, therefore, one metaethical issue of such obvious geographical interest that it cannot be by-passed: that of relativism. Its importance is heightened by the contemporary disciplinary preoccupation with difference, requiring us to 'establish ways of criticizing universalistic claims without completely surrendering to particularism' (McDowell 1995: 292), or of defending the possibility of universals against unreflective particularism. The scene may be set as follows:

> A distinctive view of value lies at the core of the cultural tradition of the West. Just as Western science has held that there are universal truths about the world, discoverable through reason and accessible in principle to people of all times and places, so Western philosophers such as Plato, Aristotle, and Kant have held that there are timeless moral truths, arising out of human nature and independent of the conventions of particular societies. This tradition acknowledges pluralism in the positive sense, but rejects it in the normative sense: it acknowledges the existence of a diversity of cultures and moral beliefs, but denies that the validity of beliefs and practices can vary across cultures. (Paul, Miller and Paul 1994: vii)

Three senses of relativism are signalled here. The first is descriptive ethical relativism, which holds that what people believe to be right or wrong differs among individuals, societies and cultures; diversity of practice exists in the moral domain. The second is normative ethical relativism, which holds that what is actually right or wrong differs among individuals, cultures and societies; what people believe to be right or wrong is right or wrong for them, in their particular context, so the validity of moral beliefs and practices can vary among cultures. The third is a form of metaethical relativism, which claims that there is no reliable way of knowing what is right or wrong, making cross-cultural evaluation impossible; some ways of life may be better or worse than others, but this cannot be demonstrated.

Descriptive ethical relativism is obviously true, to the extent that some moral values and practices are locally specific. This observation is supported by long-established facts about the world accumulated by anthropologists, sociologists and some traditional cultural geographers. Normative ethical relativism is, just as obviously, contentious, unless we subscribe to a 'vulgar' relativism (Williams 1972: 34–9), which requires equal respect for all moral codes, including those incorporating human sacrifice, institutionalised torture and so on. However, to engage the critique invited by such practices raises the metaethical question of the authority of the principles which may be invoked in describing particular moral codes as worse than others. By what right, or intellectual means, do British people condemn public floggings and beheadings elsewhere, in societies which claim that such practices work for them in their own cultural or religious tradition? How do Americans mount such a critique, when their prisoners can be kept for years in suspense on 'death row'?

Recognition within philosophy that this is an explicitly geographical issue is rare enough to cite a text using the heading of 'geographical perspective' in a section on relativism (Billington 1993: 37–8). The discussion of relativism versus absolutism (the position that there is a single moral truth) also includes the question of how judgements can be 'made geographically'. However, in referring to conflicting ideas throughout the world about how to behave, it is pointed out that one need not be an anthropologist (rather than geographer) to be aware of this. In similar vein is a description of variation in moral codes as 'a fact of anthropology' (Mackie 1977: 36).

Various strategies have been proposed to break the apparent impasse between the extremes of normative ethical relativism and absolutism, or between particularism and universalism. One is to dismiss the problem altogether, by pointing to a self-contradiction in relativism: 'the very claim that morality should be relativistic is not itself a relativistic claim. Rather, it claims to be a truth for all times and places' (Sterba 1998: 16). However, relativism (like universalism) might better be considered as a product of particular historical-geographical circumstances, making sense of different worlds (see next chapter). The strategy adopted by Seyla Benhabib (1992: 30) is therefore to situate what she describes as a 'historically self-conscious universalism' within the modern era. Her universal perspective requires that we recognise the right of all beings capable of speech and action to be participants in the 'moral conversation' within which values are debated and resolved (the principle of universal moral respect), and that each has the same rights within the conversation (the principle of egalitarian reciprocity). These form part of the presuppositions (norms and values) of the modern life-world. Benhabib (1992: 11) refers to her resolution as 'a postconventional Sittlichkeit'. The notion of *Sittlichkeit* was introduced by Friedrich Hegel as a challenge to the moral abstractions of Immanuel Kant, to refer to conventional morality as 'a concretely determined ethical existence that was expressed in the local folkways, a form of life that made particular sense to the people living it' (Williams 1985: 104). Benhabib thus reconciles moral universalism with contextual sensitivity by invoking central constituents of the moral point of view of modern (postconventional) society. Similarly, Agnes Heller (1987: 125) identifies freedom and life as the two ultimate universal values of modernity, from which other moral judgements may be justified in this particular context.

Benhabib's appeal to participation in a moral conversation as a source of moral values is a version of communicative or discourse ethics. Some ideal process of interaction with no imbalances of power among participants may be assumed, as a prerequisite for accepting the outcome of collective deliberation as right (e.g. Habermas 1990). Other attempts to seek moral truth transcending the here and now call on various propositions concerning what people would agree was right or wrong under certain conditions. These conditions might be that everyone is perfectly rational, conceptually clear, fully informed of the facts, and accepting the moral point of view in the sense of being concerned about right and wrong, or that rational reflection on relevant arguments and well-established empirical facts prevails (e.g. Miller 1992: 18). Alternatives involve the adoption of various forms of reasoning based on contractarianism (involving idealised social contracts) and constructivism (involving deduction from restricted assumptions), leading, for example, to prioritising the interests of the worst-off (Rawls 1971), the principle of reasonable rejection (Barry 1995), or the principle of non-injury (O'Neill 1996) as universal moral propositions from which specific conclusions about particular situations may be derived.

Onora O'Neill's approach incorporates the issue of the spatial scope of morality, introduced in the previous section. Her construction of practical reasoning about ethics and justice is based on the demand that 'anything that is to count as reasoning must be followable by all relevant others' (O'Neill 1996: 3). This raises the question of whether the scope is universal ('more-or-less cosmopolitan') or restricted in some way. She refers to different 'stretches' of practical reasoning in different circumstances, with borders defining insiders and outsiders. 'Connection will peter out at differing boundaries for different agents and in activity of differing sorts in differing circumstances' (O'Neill 1996: 117). The spatiality of her perspective is elaborated as follows:

> In reasoning, in justifying what we do, in criticizing what others do, we constantly appeal to a wider group . . . some thinking about and justifications of action must be presentable, hence followable and exchangeable, not merely among an immediate group of participants, or of those present, or of the like-minded, or even among fellow-citizens, but among more diverse and often more dispersed others, whose exact boundaries cannot readily be defined . . . Although the outer boundaries of *inclusive practical reasoning* cannot be set with any precision, they must be capacious in a world of multiple and diverse audiences who are linked rather than separated by porous state and regional boundaries, global telecommunications and interlocking and overlapping practices and polities. (O'Neill 1996: 53–4)

She recognises that persons cannot hope to live by spatially restricted reasoning, while stopping short of requiring its scope to be universal, and leaving open the question of how wide its scope should be in different cases. Principles must abstract from difference, but need not assume idealised accounts of human agents that deny their particularities.

Another approach with obvious geographical appeal is to accept the universality of certain grand moral sentiments or values and to recognise the spatial (and temporal) particularity of their application. Michael Walzer (1994) makes a distinction between a thin, minimalist or universal morality, captured by such values as justice and truth, and a thick, particular or local morality. This is echoed in a reference by O'Neill (1996: 68) to thick 'locally significant' categories. Taken from the notion of 'thick description' (Geertz 1973), thick means 'richly referential, culturally resonant, locked into a locally established symbolic system or network of meanings' (Walzer 1994: xi). He exemplifies a core morality differently elaborated in different cultures by pointing out that, when people took to the streets of Prague in 1989 carrying signs saying simply 'Truth' or 'Justice', everyone who saw them, wherever they were, recognised their general meaning; but what was being demanded there and then by way of truth and justice was particular to the Prague of the end of the communist era. He advocates a form of moral minimalism, which 'consists in principles and rules that are reiterated in different times and places, and that are seen to be similar even though they are expressed in different idioms and reflect different histories and different versions of the world' (Walzer 1994: 17).

Walzer's perspective is particularly helpful in pointing to a role for the geographer, in examining the contextual thickening of moral concepts in particular circumstances. Thus, the geographer takes up where the philosopher usually leaves off with the recognition that descriptive ethical relativism is so evident as to require little empirical elaboration. The geographer's traditional preoccupation with the diversity of human life can readily be applied to moral thought and practice, aware of the old axiom that the observation of differences from place to place does not deny the possibility of finding some things in common (as shown in the next chapter). The distinguishing feature of geography is therefore to approach ethical questions 'from a grounded, contextualized and often concrete perspective, which is quite unlike the style of much philosophical literature' (Proctor 1998a: 11).

Ethical deliberation is often predicated on the principle of universalisation: that persons should be treated the same in the same circumstances. This is the context in which those individual differences which preoccupy some contemporary cultural and social geographers may assume special significance. It requires careful examination of actual situations, in their full particularity, seeking to identify similarities and differences relevant to the moral issue(s) involved. This may have to do with people's structural position, like class or socio-economic status, or personal attributes like gender, sexuality, ethnicity and race. It can also involve their geographical position, or location. Understanding and practising morality in the world of actual, differentiated human nature and experience is enhanced by the investigative and imaginative capacity to know what life is like, here and now, as a basis for judgement and action.

Herein lies the most obvious and perhaps significant role for the geographer working at the interface with ethics. It is context that the geographer can help to provide. For the theories that ethicists devise, as well as the moralities people actually practise, are embedded within specific sets of social and physical relationships manifest in geographical space, reflecting the particularity of place as well as time.

In this sensitivity to context, geography shares something with two influential contemporary intellectual movements: feminism and multiculturalism. Feminist approaches to ethics have come to the fore in recent years, as challenges to what some see as a masculinist mainstream moral philosophy preoccupied with the impartial application of rights and rules. Feminist ethical theories seek to understand women's oppression, with a view to ending it, and to develop an account of morality based on women's experience (Brennan 1999). They focused initially on gender inequality and discrimination against women in patriarchal social orders, and went on to advocate a relational ethics, especially an ethic of care (see Chapter 5). Feminist moral inquiry is unlikely to proclaim any single principle applicable to all problems. More likely, according to Virginia Held (1993: 38–9), it will advocate 'a pluralistic ethics, containing some principles at an intermediate level of generality, relevant to given domains, and many particular judgements sensitively arrived at in these various domains'. She emphasises the importance of recognising the difference between contexts, but not at the expense of a relativistic ethics.

The engaged attentiveness to context and to the concrete particulars of situations, characteristic of feminist ethics, coincides with many postmodern themes (Bowden 1997: 10). In particular, feminism and postmodernism share sensitivity to difference. However, feminists have good reason to be wary of some aspects of postmodernism, resisting 'the dangerous relativism and particularism into which postmodern analyses often seem to lead' (V. Held 1993: 13), especially if relativism condones the subordination of women. Held is critical of what she sees as the postmodern abandonment of theory, and of the goal of a just society, as threatening to any marginalised group. Much more helpful are those perspectives, including strands of feminism, in which recognition of politically salient forms of difference has strengthened the hand of the oppressed. These are sometimes advanced under the banner of multiculturalism.

Two senses of multiculturalism may be identified (Okin 1998; see Chapter 6). One refers to quests for recognition and acceptance by such groups of 'others' as women, gays, disabled and 'people of colour', whose difference has hitherto been marginalised. The other refers to rights to engage in a particular culture or way of life, often involving a distinctive religion, language and history. The first sense raises obvious moral questions, including the place (in a literal, geographical sense as well as more generally) of particular gender identities and sexual practices (e.g. Binnie and Valentine 1999). The second sense raises questions of minority group rights within wider societies where different values prevail, for example the rights of Muslims in Britain to run their own schools and other community institutions. Both senses lead to the broader question of the evaluation of different ways of being, where the problem of normative ethical relativism confronts universalism. Respecting a range of different personal practices or group cultures does not necessarily entail a commitment to this form of relativism, for there may be defensible grounds for comparative evaluation. One might be the degree of tolerance and inclusion of different others. The importance of knowing the situation brings us back to a central task at the interface of geography and ethics: the conduct of inquiry sensitivity to context, as a way of dealing with difference.

THE APPROACH

It remains to outline the approach to be adopted in succeeding chapters of this book. A number of distinctive yet related strands may be identified in a geographical engagement with ethics. Perhaps the most significant is choice of subject matter itself, to focus on inequality or injustice, for example, instead of ignoring such issues. There is an obvious role for geography in descriptive ethics, observing variations from place to place in moral beliefs and practices, as part of the world of difference. Geographical interpretation may adopt a moral perspective, showing how values are implicated in the world of human creation. Solutions to normative problems may be assisted by geographical sensitivity, in the field of development for example. There is also the ethics of our professional practice. These strands will all be woven explicitly into this text, in various ways, with the exception of professional ethics. Systematic consideration of ethics in professional geography is a large enough topic to require separate treatment; the implicit ethics of this book and of its author are for the interested reader to excavate, with assistance from elsewhere (e.g D. M. Smith 1994c; 2000a).

Something of the dilemma facing geographers tempted into the borderlands of geography and ethics is expressed as follows:

> To perceive truly, which I take to mean the same as perceiving justly or morally, seems to call for the power of attending to both the particular and the universal. In our day-to-day affairs we are able to see well enough the detailed things at hand as well as their context, which is rarely large and does not have to be. If, however, we aspire to grow beyond our small world and its routines, what should we do? Is it better to attend to the things closest to us – family, friends, our own home – and move from such direct intimate experiences to the general and the universal, or is it better to start with a grasp of universal principles and values – with the distant, the grand, and the inspiring – and from that superior plane move down to concrete instances? (Tuan 1989: 177)

The interplay, and possible resolution, of two related sets of tensions provide the creative force in much of what follows. The first, already anticipated, is between the particularity and universality of morality as human practice and of ethics as attempts to comprehend this aspect of life. This involves understanding such grand abstractions as development and justice in the abstract, and also in the geographical context (cities, regions or nations) within which their specificity is expressed in actual human collaboration and struggle. Work is required in both directions: not only from the abstract to the specific, but also from the specific back to the abstractions. As Sack (1997: 230) observes: 'Conversation with real others from particular, partial, situated contexts is an essential component of applying reason and moving to a less partial position', not to the impossibility of the view from nowhere (Nagel 1986), but with enough detachment from deep embeddedness in place to set things in a wider perspective.

The second tension is between the empirical and the theoretical. In a sense, this is another expression of the relationship between the abstract and the specific. But it goes further, into the way in which observation of moral behaviour in particular contexts can contribute to the development or refinement of ethics as moral theory. Martha Nussbaum (1998: 765, 788), one of the few moral philosophers to make a sustained contribution to practical matters (in development studies), observes that 'philosophy cannot do its job well unless it is informed by fact and experience: that is why the philosopher, while neither a field-worker nor a politician, should try to get close to the reality she describes'; but she goes on to assert that we will not learn much 'if we do not bring to our fieldwork such theories of justice and human good as we have managed to work out'. Moral knowledge and ethical understanding is advanced by working back and forth, between theory and reality. 'Moral inquiry is an ongoing process . . . of continual adjustment of theory in the light of moral experience, as well as of particular judgements and actions in the light of theory' (V. Held 1993: 29). We return to this theme in the final chapter, informed by the substance of this book.

Space, place and nature are overriding metaphors informing the ontological gaze of the geographer engaging ethics (Proctor 1998a; Proctor and Smith 1999). The structure of this book reflects such familiar geographical concepts, largely overlooked by philosophers, and elaborated here so as to make something of their interlinkage in an ordered narrative. Chapter 2 introduces the historical geography of morality and ethics, exemplifying their specificity in time and space along with some similarities. Chapter 3 reviews moral readings of landscape, location and place. Chapter 4 considers the moral significance of proximity, in the context of locality and community. Chapter 5 examines the role of distance, with respect to the scope of beneficence and of some less desirable practices. Chapter 6 looks at some moral aspects of the occupation and division of geographical space. Chapter 7 discusses distributive issues subsumed under the rubric of territorial social justice. Chapter 8 moves on to moral aspects of development. Chapter 9 takes up some moral issues raised by human interaction with nature, in the context of environmental ethics. The concluding chapter considers what progress might have been made in the development of a geographically sensitive ethics.

The main title of this book, *Moral Geographies*, is taken from the most extensive and explicitly 'moral' literature in contemporary geography. Its adoption in the present context goes further, to suggest that all geographies are in some sense moral creations, reflecting various ways in which moral understandings and practices guide humankind in making life on Earth. Another way of putting it is that moral geographies are simply ethical discourses in geography (Lynn 1998: 238). However, instead of seeking definitional clarity here, the content of this book will elaborate the title, and define its field of inquiry. Let us now enter the morally-infused world of difference, which a geographical engagement with ethics might help to explain, and possibly to change for the better.

CHAPTER 2

THE HISTORICAL GEOGRAPHY OF MORALITY AND ETHICS

[A]ll knowledge is connected knowledge, located in place and time.

(Hekman 1995: 132)

Moral beliefs and practices can differ with time and place. This is the observation of moral diversity or pluralism, relativism in the descriptive sense. Ethics as moral philosophies, or the intellectual structures devised to try to make sense of the moral dimension of life, are also subject to differences, to changes over time and to variations among societies. In short, there is a historical geography of morality and ethics, the elaboration of which would demonstrate these differences and show how they arise. This chapter provides an introduction to such a project. It identifies some of the elements of the context which have a bearing on morality and ethics, provides some outlines of a historical geography and exemplifies moral diversity – along with some similarities among alternative ethical traditions.

THE SIGNIFICANCE OF CONTEXT

Morality is always contextual and historicized, even when it claims to be universal.

(Tronto 1993: 62)

Moral practice arises among persons connected together in some way, in the context of particular times and places. As practical problems of living in harmony are encountered and resolved, answers to questions of how best to behave in specific circumstances are found. With more leisurely reflection, ideas on how to live in a broader sense emerge. These may be recognised as something special, as moral theories or ethics. Their shape will be influenced by the prevailing intellectual environment, and by the material conditions in which they themselves are formed.

Obvious though all this may appear to be, the contextuality of moral practice and theory is seldom given prominence by philosophers. A notable exception is Alasdair MacIntyre, who recognises that moral philosophies 'always do articulate the morality of some particular social and cultural standpoint' (MacIntyre 1985: 268). He explains that the history of morality and of moral philosophy is one of successive challenges to some pre-existing moral order, bits of which may survive into a new era. This is rather like the way in which relics of earlier modes of production survive such economic revolutions as the rise of capitalism, or parts of previous paradigms persist after revolutions in scientific thought.

The meaning of those grand (or thin) moral values with claims to universality will be reconsidered with changing circumstances:

> Since historical conditions change continually, the question about freedom, justice and goodness has to be asked again and again. As man's concrete needs and interests change, moral norms also change and moral acts are rejudged and revalued . . . ethics consists of historical moral theories, which stand in a critical relation to each other and in terms of which continues the dialogue. (Rauche 1985: 223, 252–3)

This is the theoretical expression of the contingent character of both ethics (as theory) and morality (as practice), of the universality of humankind's contingent experience of reality, 'which causes the conception of various historical moral theories and sets in motion the interplay between theory and practice' (Rauche 1985: 253).

The particular circumstances, or forces, bearing most directly on morality and ethics will vary with the context. For example, morality may follow prevailing customs in some societies, but in others the law is recognised as codifying right and wrong behaviour. Different influences are given more or less weight in different interpretations or general social theories. Some see culture predominating, with moral values reflecting and interweaving with other aspects of a prevailing way of live. Others give primacy to economic considerations; this is the Marxian perspective which views morality as an outcome of the economic base of society and serving the interests of a ruling class. Thus according to Friedrich Engels in *Anti-Dühring,* 'moral theories are the product, in the last analysis, of the economic stage which society had reached at that particular epoch' (quoted in Peffer 1990: 271–2). The exploitation of labour is therefore consistent with the morality of capitalism, just as some persons owning others was right for those who ruled under slavery as a mode of production. Yet other interpretations stress the political dimension. For example in recognising that 'all moral theories have a *context* that determines the conditions for their relevance', Tronto (1993: 4) emphasises the power relations which impose boundaries on the shaping of moralities by excluding some views, like those of women.

Feminist philosophers have made much of this gender exclusionary character of ethics. If the place of ethical deliberation is largely the public domain, whether it be the Greek *polis*, the ecclesiastical cloister of the

Middle Ages, the assembly of tribal elders, or (until recently) the colleges of Oxbridge, it would not be surprising if those involved were predominately if not exclusively males, and from élite sections of the societies concerned. Thus, when Carol Gilligan (1982) claimed that women articulate morality 'in a different voice', stressing the importance of relationships rather than the rules predominating in mainstream ethics, she was challenging the product of a context with particular gender roles. And if we accept that moral voices are multiple and not simply those of either men or women as distinct groups (Hekman 1995), then there is a possibility that a variety of views, like those of some 'minority' groups, have been largely excluded from the discourse generating moral values and understandings.

The significance of this contextuality of ethics and morality has been recognised in geography's (re)discovery of the moral:

> The character of 'moral' discourse in our own time and place may be radically different from that of other times and places. In order to understand moral statements, one must therefore situate them historically and geographically. Different moral frameworks not only have different social and cultural consequences; they may also depend on different kinds of social and cultural constructions of things such as selfhood, society and politics. (Driver 1991: 61)

It matters much whether persons are seen as autonomous individuals, as in Western liberal thought, or so embedded within communities as to derive identity very much from being part of a group, as in premodern African tradition, for example (see Chapter 8).

A common assumption in much philosophical literature is that the prevailing perspectives of ethics are 'Western'. For example, in a review of an encyclopaedia of ethics (Becker and Becker 1992), the editor of another such guide comments that beyond 'the concentric rings of its American/English-speaking/Western core' are articles on work in other parts of the world and on historical figures 'outside the core' (P. Singer 1993: 808). His *Companion* does contain reviews of great ethical traditions – Indian, Buddhist, Classical Chinese, Jewish, Christian and Islamic – as well as ancient ethics and the ethics of small-scale societies (P. Singer 1991). Some references to non-Western ethics will appear in what follows here (and in some subsequent chapters), but for the most part this account is confined to the ethics now expressed in the English language. While this imposes a limitation on the understanding of ethics in a world of difference, the challenge of making sense even of this restricted realm is enough for one book. In any event, the underlying theme of the contextuality of morality and ethics is general enough to be applicable in non-Western ethics. Like relativism, contextualism is a universal perspective.

AN OUTLINE HISTORICAL GEOGRAPHY

> [A]ll morality is always to some degree tied to the socially local and particular.
> (MacIntyre 1985: 126)

Changes in moralities and ethics reflect changing understandings of human life, and especially of responsibility to others. Their history can be viewed as working with geography, in the sense that changes in the prevailing morality over time, and how it is understood, may be related to changing geographical circumstances, including changing knowledge of other peoples. In ancient times, as well into the modern era, localised groups had good (prudential) reasons to protect themselves from outsiders. It was hard to conceive of wider moral obligations to others, and impossible with respect to those who were both unknown and unimaginable. Ignorance could justify the exclusion from ethical consideration of those with whom the prevailing geography denied the possibility of a relationship. Pre-Columbian Europe could not be expected to have a moral stance on the treatment of American Indians, for example, nor could people polluting a river be expected to have regard for unknown others downstream. O'Neill (1996: 105) explains the cases of the inhabitants of Anglo-Saxon England and their T'ang Chinese contemporaries: 'the two groups lived in unconnected worlds, ignorant of one another's very existence: their activity assumed no connection to those in the other group, whom they accorded no ethical standing'. They also lived beyond possible inadvertent impact, in a world where trading relations tended to be transparent and where environmental degradation was locally confined rather than universalised as global warming.

A few moral philosophers have recognised the spatiality of moralities and ethics, and how they change. For example, Jeffrey Reiman (1990: 21, 158) takes principles of natural justice (e.g. the secular elements of the Ten Commandments, which prohibit killing, committing adultery, stealing, bearing false witness and coveting a neighbour's possessions) to be those that would seem reasonable to people whose social structure was simple, like a nomadic desert tribe. In contrast, he sees principles of social justice as 'those that would occur to settled territorial peoples with a complex social structure based on differential access to resources secured by a public enforcement apparatus' (Reiman 1990: 21).

The most helpful guide into a historical geography of morality and ethics is provided by MacIntyre (1985; 1998). He emphasises the importance of the historical and social context, and hence the interdependence of moral philosophy and sociology, recognising that moral concepts change as social life changes: 'Moral concepts are embodied in and are partially constitutive of forms of social life' (MacIntyre 1998: 1). The spatiality in his history of ethics is implicit, for the most part, but not difficult to tease out of his texts.

MacIntyre begins in ancient Greece, with the chieftain of Homer's epics, whose personal values of skill, courage, cunning and aggression were those required of effective leaders of men in those times. Thus 'morality and social structure are in fact one and the same in heroic society . . . Evaluative questions *are* questions of social fact' (MacIntyre 1985: 123). He explains that, with the shift from small-scale societies, these very values came to be viewed as anti-social, noting 'the difference that it makes to the conception of the virtues when the primary moral community is no longer the kinship group, but the city-state' (MacIntyre 1985: 132). As one set of social and spatial forms replaced another, so the most valued characteristics of men changed (while women maintained the same invisibility).

Homer's characters had no way to view their own culture and society from any distance, as if from the outside. However, the process of invasion and colonisation increased trade and travel, bringing home the fact of different cultures: 'the distinction between what holds good in Egypt but not in Persia, or in Athens but not in Megara' (MacIntyre 1998: 10). Traditional role behaviour could then be seen in contrast with other possibilities, and choice between old and new ways became a fact of social life. He goes on: 'it is not surprising therefore that it is in the transition from the society which was the bearer of the Homeric poems to the society of the fifth-century city-state that *good* and its cognates acquired a variety of uses, and that it is in the following decades that men [*sic*] reflect self-consciously about those uses' (MacIntyre 1998: 84).

In the fifth century BC each city-state had its own conventions, dominant practices and criteria of justice. However, the differences observed among cities with respect to customs and laws raised the problem of whether justice should differ from city to city, whether justice holds only within a given community between its citizens, or whether it should hold also between cities. Hence the geographical question, as put by MacIntyre (1998: 16): 'Where and how shall I choose to live?' In the absence of transcending (trans-city) values, there was no way of making this decision, other than on the basis of personal preference. Normative as well as descriptive ethical relativism prevailed.

However, Plato placed the actual forms of constitution of Greek cities on a moral scale. So, even if we cannot have the ideal, we are told that timocracy – a form of government in which there is a property qualification for office – (in traditional Sparta) is best, oligarchy (in Corinth) and democracy (in Athens) worse, and tyranny (in Syracuse) worst of all

(MacIntyre 1998: 45–6). This was because a type of personality corresponds with each type of constitution: in timocracy the human appetites are restrained and ordered, in oligarchy and democracy they are less disciplined, while in tyranny the baser appetites are in control. Some conception of human nature thus generated the comparative evaluation of different forms of city life.

A theory of human nature was also implicated in the particular virtues which defined the good life in the Greek city-state or *polis*. These were dispositions of character with both instrumental and intrinsic value to their owners, and wholly explicable in terms of the good for man and requiring no external justification (Brown 1986: 148). The list of virtues in Aristotle's *Ethics* reflects what he takes to be the code of a gentlemen in contemporary Greek society: courage, temperance, liberality, magnificence, greatness of soul, good temper or gentleness, being agreeable in company, wittiness and modesty (MacIntyre 1998: 67–8). This list has two parts: traits such as courage and agreeableness which it is hard to conceive of as not being valued in any human community, and those which it would be hard to comprehend outside Aristotle's own social context and preferences. The development of those personal dispositions, qualities or virtues relevant to the preservation of morality in this particular community, with its shared conception of the good life, depended on highly localised spatial interaction: 'individuals need to be fairly close to one another in order that they can observe, correct, direct, and help to shape the dispositions of themselves and others' (Tronto 1993: 30).

Another aspect of the spatiality of the ethics of the Greek city-state was manifest in its collective provision for the general welfare, in the form of defence against evident external threat, while the leaders appeared indifferent to the poverty which afflicted some of those within (Walzer 1983: 69–71). And of course the slavery of non-Greeks was part of the way of life, as was the exclusion of women from the public sphere, from the virtues associated with it and from the discourse sustaining the prevailing moral and social order.

In passing, it is worth noting a similar process of change in ethics elsewhere, in a shift from pre-Islamic Arab culture bound primarily by local oral tradition, to one based on a revealed text. In his book *The Virtuous City* (writen in the ninth century), Al-Fabari argued for citizens acquiring traits to enable them to function as residents of a virtuous *polis*, suggesting a communal framework for attaining ultimate happiness (Nanji 1991: 111).

The eventual decline of the Greek *polis* and the rise of the large-scale state had important consequences for the history of moral philosophy:

> The milieu of the moral life is transformed; it now becomes a matter not of the evaluations of men living in the forms of immediate community in which the interrelated character of moral and political evaluation is a matter of daily experience, but of the evaluations of men often governed from far off, living private lives in communities which are politically powerless . . . the individual finds his moral environment in his place in the universe rather than in any social or political framework. (MacIntyre 1998: 100)

The large impersonal kingdoms and empires of the Hellenistic and Roman worlds placed the individual in the context of a cosmos, not of a local community. Whereas in modern urban life the state's repressive power is close at hand, in these earlier periods of the spatially extensive state it was often absent or far away (MacIntyre 1998: 134).

Idealisation of the city-state of ancient Greece (and, later, of Renaissance Italy) has much to do with its spatial scale, which is seen to encourage a particular kind of moral life:

> Its human dimensions beckon us still. Small enough so that each person would know his neighbour and could play his part in the governance of the city, large enough so that the city could feed itself and defend itself; a place of intimate bonding in which the private sphere of the home and family and the public sphere of civic democracy would be but one easy step apart; a community of equals in which each would have enough and no one would want more than enough; a co-operative venture in which work would be a form of collaboration among equals.
>
> (Ignatieff 1984: 107)

This is reflected in Rousseau's vision, which involved restricting the scale of republics so that identity of interests could be clear to all citizens; his utopia is 'a republic of needs, a society which by democratically limiting its size, its contact with the outside world, and most of all the domestic consumption of its citizens, reduces inequality, envy and competition' (Ignatieff 1984: 122–3).

With the rise of Christianity, the emergence of long-lasting religions involved 'a set of beliefs and ways of behaving which become relatively independent of particular, specific forms of social life'; hence the expectation of 'a great capacity for coming to terms with quite different sets of moral standards in different times and places' (MacIntyre 1998: 110–11). The individual and the state became separated, with persons looking to transcendent criteria rather than to those implicit in the practices of a particular political community such as the *polis*. The increasing complexity and spatial scope of social life required a level of mutuality which governments could not be relied on to provide (Preston 1991: 95). Hence the shift from social identity in community to individual choice, or moral sovereignty, increasingly detached from particular social roles but guided by extra-terrestrial authority.

Tuan suggests roughly contemporary changes in moral philosophy in China, with the widening spatial scope of society. A central teaching of Confucius (c. 500 BC) was that the family, with its filial piety and kinship obligations, provides a model for society as a whole. However, effective rule of an empire was found to require laws because humans have 'a natural tendency to favor their own kind, which in practice often means taking unfair advantage of those who lie outside the fold' (Tuan 1989: 45). The Confucian model was therefore enlarged to cover the entire empire, with the emperor as paternal figure for all the people and lesser paternal figures for smaller groups in a nested hierarchical structure: 'the natural affection that resonated primarily among kinsfolk . . . could reach out in diminishing degrees to neighbors and strangers' (Tuan 1989: 117–18).

We now skip many centuries, to the dawn of the modern era. Up to this time, in most parts of the world, a person's identity, and morality, was closely bound up with membership of some localised group: 'Individuals inherit a particular space within an interlocking set of social relationships; lacking that space, they are nobody, or at least a stranger or an outcast' (MacIntyre 1985: 33–4). Thus:

> [I]n much of the ancient and medieval worlds, as in many other premodern societies, the individual is identified and constituted in and through certain of his or her roles, those roles which bind the individual to the communities in and through which alone specifically human goods are to be attained. (MacIntyre 1985: 172)

With the breakdown of feudalism in Britain, for example, more autonomous persons began to replace their deeply embedded ancestors. Freedom had a particular meaning, in the context of release from feudal obligation and inherited roles:

> What freedom is in each time and place is defined by the specific limitations of that time and place and by the characteristic goals of that time and place . . . when we speak of men as being unfree, what we mean is always relative to an implicit normative picture of human life, by means of which we identify what human bondage is. (MacIntyre 1998: 204)

Thus, circumstances alter virtues: a central theme of MacIntyre's historical and social account of morality and ethics.

Rights of citizenship augmented the traditional virtues or duties as a locus of morality. That these rights could be given substance was itself a product of a new form of spatial political organisation: the sovereign nation-state. Rights themselves depend on the law, on universality in the sense of 'the exceptionless rule of one set of laws on the territory over which their sovereignty extended' (Bauman 1993: 8), backed up by the coercive power of a state with the same spatial scope. But as MacIntyre

(1985: 70) explains, while the concept of rights was generated to serve one set of purposes in connection with the autonomous moral agent, another concept, that of utility, was devised for another set of purposes: those of accounting for economic behaviour (construed as individual satisfaction seeking). This set up the possibility of moral conflicts: between claims invoking rights on the one hand and utility on the other, and between these and claims based on some traditional concept of justice, for which there was no rational resolution – no way of prioritising or weighting one claim or criterion over another. Thus the problem of moral incommensurability, which continues to trouble us to this day, 'is itself the product of a particular historical conjunction'.

It was towards universalism that the sentiments of modernity were moving. Joan Tronto (1993) recognises the geographical context in which changing forms of life required people to think differently about morality. She explains the relationship between the emergence of universalistic morality and the expanding spatial scope of eighteenth-century life. In earlier societies typified by the Greek *polis*, members of the community shared some conception of the good life, and lived close enough to observe and correct behaviour. Now, people had to devise ways of living with much more distant others, with whom there might be little intimacy, in 'a geographically large, diverse, market-oriented world' (Tronto 1993: 29). She elaborates:

> In the eighteenth century, literate Europeans saw themselves as part of an increasingly broader society whose social, moral, and political concerns were increasingly less parochial and more universal . . . While humans grew more distant from one another and the bonds between them became more formal and more formally equal, they also had to expand their gaze beyond the local to the national, and indeed sometimes to a global level . . . these ideas required a change in the nature of moral thought from a type of contextual morality to a morality where human reason could be presumed to be universal. (Tronto 1993: 31–2)

These thinkers believed that they had found a way of coping with the many diverse peoples and ways of life which were being encountered. Conspicuous though some of the differences might have appeared, in the frequently exaggerated accounts of explorers and merchants, as distant peoples became more familiar a greater sense of common humanity developed. However much the treatment of these peoples, on the part of empire builders for example, departed in practice from the impartial ideal associated with the Enlightenment, the ascendance of moral universalism in theory was the outcome of historical-geographical circumstances. As Tuan (1989: 169–70) observes, enlarging the sphere of moral concern to embrace the entire world is one of the 'changes in the moral landscape' achieved by the moral imagination of the modern period. Thus, moral universalism is itself contextual, geographically as well as 'a historical result' (Habermas 1990: 208).

Tronto's sensitivity to spatial relationships and scale in the changing emphasis of morality is further illustrated as follows :

> As the bonds with the familiar grew more attenuated, connections with those who were more distant grew more prominent, if not necessarily more strongly felt. Within this *new spatial order*, the question of how much moral values could draw upon the familiar became central. By the end of the eighteenth century, moral theories [*sic*] that drew upon the local for its logic, its creation, and its expression, were no longer viable. (Tronto 1993: 38, emphasis added)

Emerging in their place was more of a rule-bound morality. Individuals following rules do not need to know much about other individuals, presumed to be following the same rules: a universalistic morality does not have to assume intimacy among members of the same moral community, who may be located great distances from one another (Tronto 1993: 29).

An important consequence was a growing tension between the notions of moral sentiment and sympathy associated with the partiality of the parochial and particular social relations of the premodern era, on the one hand, and the universal sense of benevolence associated with more impartial relations, on the other. Tronto (1993: 41) highlights the role of distance in this tension, quoting from the work of the contemporary Scottish Enlightenment philosopher Francis Hutcheson: 'This *universal Benevolence* toward all men, we may compare to that Principle of *Gravitation*, which perhaps extends to all Bodys in the *Universe*; but the *Love* of *Benevolence*, *increases* as the distance is diminish'd'. This prescient anticipation of the gravity model revered by the spatial science school of human geography pointed to the problem of extending the spatial scope of moral sentiments (to be explored more fully in Chapter 5). Another contemporary, David Hume, recognised that, while benevolence is at the root of justice, benevolence alone would not always extend far enough to create justice: as an artificial virtue, justice would depend on the natural virtue of benevolence, reinforced by human convention and law (Tronto 1993: 45). However, the triumph of universal justice over sympathy, or reason over feeling, meant that moral sentiments were increasingly 'displaced' (literally) from moral life, increasingly 'located within the private household' (Tronto 1993: 55).

All this had important implications for moral theory. The decline of the household as the prime unit of economic production, as the factory system eroded domestic industry, tended to separate the domestic and public spheres of life further. The domestic sphere, associated with women, became the repository of those moral sentiments which were being replaced by a more impartial, universal, rule-bound morality characteristic of public affairs dominated by men. This is how Tronto explains the way in which the historical and geographical circumstances of eighteenth-century life contained both women and moral sentiments within the domestic sphere. And there they stayed, for the most part, until different circumstances encouraged the (re)emergence of a women's perspective in the 1980s, in the form of an ethic of care which challenged the hegemony of an ethic of justice (see Chapter 5).

An important geographical force in moral change was urbanisation. The rural-to-urban population shift characteristic of the eighteenth and nineteenth centuries is often associated with a move from what the German sociologist Ferdinand Tonnies termed *Gemeinschaft* (a community built around kinship, attachment to place and cooperative action) to *Gesellschaft* (a society of impersonal relations founded on formal contract and exchange): from a communitarian identity and morality of partiality to an autonomous identity and impartial morality. This had specific consequences, in the formation of new social relationships. One was the rise of public welfare services in cities; as Tuan (1986: 76, 78) observes:

> What distinguishes the city is the scope of its aid to strangers . . . Charity in the past and welfare in the present arose in response to urgent needs of the urban populace [involving] the workings of a moral imperative, of an impulse toward impersonal justice, a desire to give aid to all those who are in need.

Another was the tendency of geographical mobility to 'liquidate conventional notions of hierarchical authority', including patriarchy; women in particular found a new egalitarian ethic in the towns, and some were able to avoid aspects of patriarchal control – if not economic exploitation (Turner 1986: 123).

So, increasing geographical mobility became a crucial element in the changing context of moral life. Indeed, the history of modernity might be read as a history of the movement of European peoples out of confining communal place into the open space of modern society. The celebration of individual autonomy and liberty was closely associated with release from a socially assigned place in a spatial and structural moral order, and with freedom to find one's place rather than to know one's place. For some persons, this might merely represent substituting one parochial situation for another: the farming community or immigrant neighbourhood in America for the European village, for example. But for many it involved opening up experience of the world of difference, and an increasingly cosmopolitan or international perspective.

This is part of the inheritance of the present era of globalisation. Something of the responsibility of the full modern citizen is captured as follows:

> Definitively torn from localism or regionalism by the modern nation-state, our modern citizen/ethical subject, already the product of a complex historical process, was then thrown into world history. Although the process has taken place unevenly, we have all come to live economically, ecologically, politically, and militarily in one interdependent world. (Aronson 1990: 71)

Onora O'Neill helps to reveal some of the spatiality of moral change in this context. In early modern and subsequent times a form of universalism with expanded scope was required to replace particularism, in a world where local communities were being incorporated into modern states, and Europeans were colonising the lands of others (O'Neill 1996: 28–9). To ground justice in appeals to particular practices and traditions would have frustrated the process of imperialism. It would also have challenged the rectitude of the project of modernisation, or the diffusion of western values, which became central to the subsequent discourse of development (see Chapter 8).

As uneven development became a moral issue, towards the middle of the present century, attitudes changed. International distributive justice was added to the moral agenda. As O'Neill (1991: 276) explains, thinking in terms of global distribution is a fairly new possibility. With the great empires of the past, their boundaries were also the boundaries of redistribution; global distributive justice was hardly imaginable. Indeed, until recently the spatial scope of the discourse of social justice was taken (usually implicitly) to be the nation state. That we can now think on a broader scale is in part the result of a changing geography: 'the modern world is different from its predecessors. It is not a world of closed communities with mutually impenetrable ways of thought, self-sufficient economies and ideally sovereign states' (O'Neill 1991: 282): hence, among other outcomes, the institution of the United Nations and its commitment to universal human rights, as what may come to be seen as a final flourish of modern universalist idealism before a new age of parochialism set in.

The upsurge of communitarian thinking in the 1980s, against the reality of economic and political structures becoming more cosmopolitan, could be interpreted as follows:

> Might it reflect the fact that cosmopolitan claims are no longer advantageous to elites, as they perhaps were or were thought to be in the recently past era of imperialism? In a post-imperial world, cosmopolitan arrangements threaten rich states with uncontrolled economic forces and immigration and demands for aid for the poor of the world, and autocratic states with demands that human rights be guaranteed across boundaries. (O'Neill 1996: 28–9, note 31)

Extending this line of argument, it could be that the rise of postmodern thinking, with its disdain for universalism and encouragement of ethical relativism (or nihilism), undermines the scope for critique of an increasingly unequal world, very much to the disadvantage of those marginalised peoples whom some versions of postmodernism claim to empower. Thus:

> Postmodernism has indeed forced us to focus on the local rather than the international, thus obscuring the extent to which today the local itself is but an extension of the international. My sense is that as the social reality which we face gets increasingly more global, complicated and intricate, our units of analysis get increasingly smaller, fragmented and marginal. (Benhabib 1992: 241, note 69)

A return to the premodern myopia of concern with the local and particular, in virulent nationalisms for example, may be an understandable political reaction to globalisation, the benefits from which seem increasingly local and particular. But it is hard to see this as moral progress.

Zygmunt Bauman's work on postmodern ethics further illustrates recognition of the spatiality as well as the temporality of morality. He points out that the moral agenda of our times is full of items which writers of the past hardly touched, if at all: 'they were not articulated then as part of human experience' (Bauman 1993: 1). He mentions issues arising from the plight of pair relationships, sexuality and the family, and the multitude of traditions vying for loyalty and authority, as well as the global context of contemporary life. These place new demands on ethics. His example of contemporary warfare is an especially vivid illustration of the moral implications of changing geo-technical relations:

> With electronic surveillance and smart missiles, people can now be killed before they have a chance to respond; killed at a distance at which the killer does not see the victims and no more has to (or, indeed, can) count the bodies. (Bauman 1993: 227)

Even more remote killing is associated with environmental change; for example, millions of lives are at risk from drought and floods arising from global warming caused by the so-called greenhouse gasses released largely in the industrialised parts of the world.

Bauman and others influenced by postmodern thinking signal a radical challenge to modes of ethical thought inherited from Enlightenment universalism. For example Linda Nicholson (1990: 1–2) criticises feminist theory from the late 1960s to the mid-1980s for a tendency to reflect the viewpoint of white, middle-class women of North America and Western Europe; she sees this as representing errors of universalism carried over into feminism, failing to recognise 'the embeddedness of its own assumptions within a specific historical context . . . not used to acknowledging that the premises from which they were working possessed a specific location'. While her usage of 'location' is more general than the geographical sense, an America-European perspective would exclude women of the Third World. However, the postmodernism with which she engages is not simply a matter of recognising new moral issues in a new context requiring new ethics. Postmodernism claims to go beyond historicist interpretations of the situatedness of human thought (like moral values) within cultures, to question claims about how knowledge is legitimated through criteria of truth. Postmodernism thus offers 'a wariness towards generalisations which transcend the boundaries of culture and region' (Nicholson 1990: 5).

So, what do postmodern perspectives on ethics actually provide? The very term 'postmodern' signals some kind of change, something after modern thought. This involves at least a suspicion of grand theory, and of claims to truth which transcend the context of time and space. But if it is literally postmodern, it cannot merely endorse the normative ethical relativism associated with the premodern era. Bauman (1993: 4) recognises some continuity with the modern: 'The great issues of ethics – like human rights, social justice, balance between peaceful co-operation and personal self-assertion, synchronization of individual conduct and collective responsibility – have lost nothing of their topicality; they only need to be treated in a novel way'. But his assertion that we need to abandon universalism and give primacy to the individual moral conscience, based on feeling rather than reason, and operate case by case rather than obeying general rules, has been countered by a reminder that these values are characteristic of simpler premodern societies (Matthews 1994). Another sceptic, provoked by deconstruction in literary criticism, goes even further back: 'So-called postmodern thinking is much better described as pre-ancient (pre-Socratic, pre-Aristotelian) thinking. Or rather, more or less developed versions of these ideas surrounding these issues have been around in *every* age, with more or less popularity' (Matchen 1995: 93–4, note 13). Could they even have an element of universality?

The purpose here is not to engage a critique of postmodernism. It is simply to show that the very time-specificity of what is claimed to be postmodern thinking, in ethics and more generally, is itself contested. What is clear, nevertheless, is that some knowledge of geographical as well as temporal context is necessary to any understanding of change in morality and ethics, whether construed as progress, regress, or merely difference. The glimpses provided here into the ways in which some moral philosophers have sought connections between moralities theorised or actually practised and the spatial relations in which they are embedded provide pointers to how a fuller historical geography of ethics might be constructed.

MORAL DIFFERENCES AND SIMILARITIES

[M]orality is local. There are no *interesting* moral universals. (Posner 1998: 1640)

[W]hen comparing cultural traditions, one is struck as much by the similarities as by the differences. (Kukathas 1994: 10)

There is widespread recognition that moral diversity, or pluralism, is a fact. This resonates with the contemporary resurrection of difference in human geography, adopted well beyond those who explicitly embrace postmodernism. Indeed, respect for difference has almost become emblematic of an academic and political stance sensitive to multiculturalism and hostile to the essentialism associated with oppressive assertions of a general human nature. Yet, there are also those who stress similarities among cultures, their moral codes and understandings.

Whether an emphasis on differences or similarities is appropriate depends very much on the kind of values or behaviour under consideration. For example, there may be widespread if not universal inter-cultural agreement on fundamental values, such as persons not harming others, but less agreement on more specific activities such as factories not fouling the air, and less again on personal behaviour like not smoking in public, even though accepting the fundamental value appears to entail restraint on both business and individual polluting practices. A Muslim and a Christian may agree on monotheism (though disagree on which is the true god), but disagree on the role of women in society and how they should be treated (Kukathas 1994: 11). The meaning of such values as honour and disgrace is highly dependent on the customs and practices of particular times and places, while the natural experience of pain and pleasure may be more universal (Graham 1990: 30).

Relativism in the descriptive sense may best be demonstrated by pointing to particular important kinds of differences in moral beliefs, which deny the existence of a single true morality. To one writer on relativism, a striking candidate would be the emphasis on individual rights embodied in the ethical culture of the modern West, absent in traditional cultures of Africa, China, Japan and India where duties are organised around the common good of a certain ideal community life (Wong 1993: 445). However, he goes on to suggest that, rather than accepting any practice, a 'substantial' or normative relativism would rule out moralities which aggravate interpersonal conflict, and would recognise that 'adequate' moralities must involve persons capable of considering the interests of others. That morality is essentially an other-regarding practice is commonly espoused as universal truth.

But, is this notion of morality, or the very idea of morality, itself historically (and geographically) relative? Is it specific to fairly sophisticated cultures, with the capacity to discuss values, and to register them in some way necessary to their reproduction? Morality in 'simple' societies may be by no means simple, but articulated and understood in a way which differs significantly from modern Western intellectualism:

> A people whose material culture is simple may nevertheless have acquired complex rules of social behavior . . . such a people's oral tradition may contain tales of a richly wrought and highly imaginative kind. However, skill in language does not necessarily mean that it will be used to build a moral edifice – an integrated view of reality which attempts to understand the things that befall humankind as well as delineate in a fairly systematic way how people ought to behave. The Mbuti [African hunter-gatherers] and the Semang [of Malaya] do not see the need for reflections on moral issues; like many other nonliterate peoples they have not sought solace in an elaborated moral-ethical system. (Tuan 1989: 28)

The distinction appears to be between practising morality and creating ethics. Relationships with others are at the heart of morality, but conceptions of how this ought to be differ among societies in coherence and comprehensiveness, subtlety and depth. Tuan (1993: 241) suggests that the scale of a moral system tends to vary with the size of the group: those of hunting bands are likely to be smaller, less fully developed than those of large sedentary groups.

Two cases may be summarised briefly, taken from an anthropological collection mentioned in Chapter 1 (Howell 1997). One is set in Mongolia and the other in Africa; both reveal fundamental departures from Western understandings of morality. Caroline Humphrey (1997) points out that there is no single term in the Mongolian language that corresponds, however loosely, to the European concept of morality, while Anita Jacobson-Widding (1997) also fails to find among the Shona-speaking people of Zimbabwe local words that can be satisfactorily translated into 'moral' or 'morality'.

The case of Mongolia illustrates the practice of persons using an ideal exemplar, such as a heroic historical figure, as a guide to behaviour in particular circumstances. The only similarity in Western ethics is the identification of 'moral saints' (e.g. Hinman 1994: 266–72), like Mahatma Ghandi and Mother Teresa. The ethics of exemplars differs from the practice of following rules more typical of the West. It requires the subject to do some work, to ponder the meaning of the exemplar, in a moral discourse which is open-ended and unfinished (Humphrey 1997: 34). Another difference is as follows: 'for Mongols the core of morality is primarily referred to the self, adjudicating one's own actions as good or bad *for oneself*, whereas in the West at the very least a sympathy for others has been considered by most recent philosophers as a sine qua non for entering the world of morality' (Humphrey 1997: 32–3). Thus, while the practice of personal ethical egoism is recognised in Western ethics, it is not usually accorded much interest, or respect.

Another kind of departure from the expectations of Western ethics is examined by Jacobsen-Widding (1997). The three sins of lying, farting and stealing featured in her title, taken from a Fulani proverb, exemplify lack of self-control. The association is with shame rather than guilt: they are sins of commission, rather than of omission in the sense of failing to meet obligations or duties. This is in contrast to the author's definition of morality, heavily dependent on her own culture, with its emphasis on individual responsibility, conscience and guilt. To show self-mastery is a way of recognising the relative position of people who interact, how the Fulani define themselves in relation to others, to demonstrate 'Fulaniness' rather than individual identities. It expresses collective social personhood, 'what it is to be a person among other persons' (Jacobsen-Widding 1997: 51), a version of the common African communitarian ethic of 'a person is a person through persons' (Shutte 1993: 46).

These two cases could be multiplied. For example, there are evident differences among cultures in sexual morality, the treatment of women, and kinship obligations. However, they should be sufficient to illustrate the fact of diversity in moral values and practices.

Turning to similarities, the scene may be set as follows:

> [E]thics is *not* a meaningless series of different things to different people in different times and places. Rather, against a background of historically and culturally diverse approaches to the question of how we ought to live, the degree of convergence is striking. Human nature has its constants and there are only a limited number of ways in which human beings can live together and flourish . . . Hence what is recognized as a virtue in one society or religious tradition is very likely to be recognized as a virtue in the others; certainly, the set of virtues praised in one major tradition never make up a substantial part of the set of vices of another major tradition. (P. Singer 1991: 543–4)

If there is a prime candidate for the status of a universal, trans-cultural norm, understood and practised in a similar way irrespective of time and place, it is reciprocity. 'Among the characteristics common to all moralities . . . it seems that sociability is a universal human trait and reciprocity appears to be a functional necessity of sustained relationships' (Silberbauer 1991: 27). The only possible exceptions are societies subject to serious disruption, for example, where former neighbours turn on one another (as in the Balkans), exceptions which themselves underline the norm. Reciprocity is reflected in the everyday practices of mutual aid, exchanging gifts, returning favours and taking turns, and in such axioms as 'I'll scratch your back if you will scratch mine', as well as in more formal social rituals or routines. Denial of the capacity to participate in such reciprocal social relations may be tantamount to exclusion from the prevailing moral community. For example, one of the most depriving features of poverty in the (former Soviet) Republic of Georgia is inability to participate in the life of friends and relatives by giving presents and providing material aid, thus detaching people from their traditional social environment (Tarkhan-Mouravi 1998: 96). There is more to a Georgian feast than the food: as elsewhere, mutual hospitality binds persons into community membership.

The principle of reciprocity has been generalised in what is usually referred to as the Golden Rule. Treating others as you want to be treated yourself is almost universally accepted across the world as a rough-and-ready guide to life with others. Preston (1991: 95) refers to the 'natural morality' of the Golden Rule in Christian ethics: 'Always treat others as you would like them to treat you' (*Matthew* 7: 12). It is found in a similar form in other traditions, for example in Hinduism, Buddhism, Confucianism and Judaism, as well as in ancient Greek writings (Reiman 1990: 147). Confucius formulated a negative version: 'What you do not desire, do not effect on others' (Hansen 1991: 72). In Jewish ethics, the famous rule of the rabbinical authority Hillel, enunciating what he would tell a non-Jew who asked to be taught the entire Torah while standing on one foot, was: 'What you dislike don't do to others; that is the whole Torah' (Kellner 1991: 86-7). The Golden Rule expressed as 'love your neighbour as yourself' is based on the essential equality of all human beings: neighbour must be understood as not merely equal but of equal worth in the most exalted sense of being made in the image of God (Goldberg and Rayner 1989: 298).

The near universality of the Golden Rule strongly suggests that it may correspond to 'a natural tendency of reason, perhaps the very structure of conscience itself' (Reiman 1990: 147-8). If it is close to being a global principle, with roots in a wide range of cultures, the Golden Rule may provide a standard to which different peoples can appeal in resolving conflicts (H. J. Gensler 1998: 112). It is incorporated into the formal apparatus of Western moral philosophy, as the principle of universalisability:

> Universalizability enjoins us to reverse perspectives among members of a 'moral community' and judge from the point of view of the other(s). Such reversibility is essential to the ties of reciprocity that bind human communities together. All human communities define some 'significant others' in relation to whom reversibility and reciprocity must be exercised – be they members of my kin group, my tribe, my city-state, my nation, my co-religionists. What distinguishes 'modern' from 'premodern' ethical theories is the assumption of the former that the moral community is coextensive with all beings capable of speech and action, and potentially with all of humanity. (Benhabib 1992: 32)

This may be as close as everyday human experience and ethical reflection get to converging on the recognition of a moral perspective with claims to universality.

Another way of approaching moral universality is to invoke the avoidance of harm or injury to others (e.g O'Neill 1996). Indeed, some accounts of reciprocity specifically stress harm as well as help; for example, 'people should help those who have helped them; people should not injure those who have helped them' (Selznick 1992: 97). What constitutes harm seems to be universally understood; thus,

> [E]very society falls back on a quite limited range of punishments such as deprivation of money or property, physical confinement, loss of bodily parts, pain, and death. Unless these were regarded by people with a wide variety of conceptions of the good as evils, they would not function reliably as punishments. (Barry 1995: 141)

The general human experience of well-being and harm point to further possibilities of a trans-cultural moral perspective with universal tendencies:

> [B]asic codes of moral conduct and deliberations about what may or may not be rightly done relate in definite, if not always direct, ways to considerations of suffering – the avoidance and alleviation of it – and to the promotion and maintenance of well-being, in the largest sense. This is the type of consideration which nearly anyone can understand as the possible candidate for a compelling, action-guiding reason, because in the experience of everyone will be some kind of knowledge, albeit of variable breadth, depth and sensitivity, of what suffering and well-being actually feel like. (Geras 1995: 94)

In similar vein, Gilligan (1987: 20) invokes the generality of two other values:

> Since everyone is vulnerable both to oppression and to abandonment, two moral visions – one of justice and one of care – recur in human experience. The moral injunctions, not to act unfairly toward others, and not to turn away from someone in need, capture these different concerns.

That justice and care may themselves converge on a special responsibility to society's weakest, most vulnerable or needy members is another strand of moral thought connecting different traditions. Goldberg and Rayner (1989: 301) point to the 'characteristic Jewish emphasis that the rich, in giving to the poor, are not doing themselves a favour but discharging an obligation they owe as a matter of justice'. Similar is the Quran's emphasis on the ethics of redressing injustice in economic and social life, with individuals urged to spend their wealth on particular categories of the needy (Nanji 1991: 108). Thus, these and other religious traditions invoke an ethic of sharing individual and communal resources with the less privileged.

If justice and care may be added to the reciprocity of helping and not harming, as candidates for something approaching universality in moral values, others surely follow. For example, most cultures have similar norms against killing, stealing and lying (H. J. Gensler 1998: 16). As the list is extended, two issues emerge, or re-emerge. The first is the basis of such common values, if such they are: how do they arise, what justification do they have? The second is the significance of differences in their application, from place to place as well as time to time, the challenge of relativism (once again) to claims of universalism.

One approach to the first issue is that of realism, which assumes that there is some moral truth to be discovered. A possible starting point is similarity in the human situation:

> A moral dilemma or issue faced by one person is a dilemma or issue that might be faced by another ... If agents in the same circumstances act in the same way either they both act rightly or they both act wrongly. (M. Smith 1994)

This suggests the possibility of a convergence of moral opinion which might be considered objective. Alongside some entrenched disagreements are massive areas of agreement. This is the significance of those moral concepts, that both describe some naturalistic state of affairs and positively or negatively evaluate it (Williams 1985: 129), like courage, brutality, honesty, duplicity, loyalty, meanness, kindness and treachery:

> [W]hat the prevalence of such concepts suggests is that there is in fact considerable agreement about what is right and wrong . . . moral agreement is in fact so extensive that our language has developed in such a way as to build an evaluative component into certain naturalistic concepts. (M. Smith 1994: 188)

A more direct argument from naturalism is exemplified by Philip Selznick (1992). Ethical naturalism brings knowledge of a natural order to bear on questions of moral truth, such that biological and psychological as well as social conditions affect moral ideals. Thus:

> Friendship, responsibility, leadership, love, and justice are not elements of an external ethic brought to the world like Promethean fire. They are generated by mundane needs, practical opportunities, and felt satisfactions . . . authentic human ideals have material foundations. They are rooted in existential needs and strivings . . . genuine values emerge from experience; they are discovered, not imposed.
> (Selznick 1992: 19)

Morality thus has its roots in the facts of human dependency and connectedness, in the inescapably social nature of human existence. He proposes that a long list of moral universals could be drawn up, including the fact of morality itself (involving subordination of individual inclinations to the welfare of the group), preserving human life, looking to the well-being of close relatives, prohibiting murder and theft, and valuing affection and companionship, as well as reciprocity in helping and being helped. These may be expressed differently in different societies, but their universality shows that human societies are everywhere much the same in their appreciation of basic morality: 'it is just as easy to be impressed by the uniformity as by the variation' (Selznick 1992: 96). Thus: 'Appreciation for diversity and particularity is no bar to drawing general conclusions about moral experience . . . moral experience is fluid and variable, but it is also the crucible within which enduring values are established and institutional learning takes place' (Selznick 1992: 36, 37).

Selznick (1992: 106) recognises that every system of thought may contains a local element – some idea or perception that is wholly indigenous and idiosyncratic, but that 'the real question is, what difference does it make?' As an ethics text puts it: 'Many moral differences can be explained as the application of similar basic values to differing situations' (H. J. Gensler 1998: 16). And these situations will include those of geography. Again, we encounter the question of the spatial scope of values with claims to universality. There are shifting groups with changing moral conventions: 'Different conventions are appropriate in one's family, in the local neighborhood, with co-workers, with friends, and with strangers in one's society' (Harman and Thompson 1996: 7). Different attitudes to insiders and outsiders may prevail; for example a person who is able successfully to cheat outsiders may be admired. Such is the morality of locals fleecing tourists, and of honour among thieves. But their significance is as departures from widespread convention, rather than as alternative and competing moralities. Cheating and stealing cannot be universalised as a basis for social life which is sustainable, never mind admirable.

We conclude by returning to the work of Alasdair MacIntyre. He traces back to Aristotle the proposition that 'there are certain features of human life which are necessarily or almost inevitably the same in all societies, and that, as a consequence of this, there are certain evaluative truths which cannot be escaped' (MacIntyre 1998: 95). Thus, 'the recognition of a norm of truth telling and a virtue of honesty seems written into the concept of a society', which could hardly function without them; other virtues are causally necessary to the maintenance of social life, given widespread and elementary facts about human life and its environment, so 'the existence of material scarcity, of physical danger, and of competitive aspirations bring both courage and justice or fairness on the scene' (MacIntyre 1998: 77). The significance of these common virtues is summarised as follows: 'truthfulness, justice and courage – and perhaps some others – are genuine excellences, are virtues in the light of which we have to characterise ourselves and others, whatever our private moral standpoint or our society's particular codes may be', but this is 'compatible with the acknowledgement that different societies have and have had different codes of truthfulness, justice and courage' (MacIntyre 1985: 192).

So: 'Common standards, common ideals, common tastes, common priorities that make a common morality possible, rest on shared joys and sorrows and all require active sympathy' (Midgley 1991: 12). The kind of consideration for others generalised by the Golden Rule is central to the mutual understanding which arises among naturally social beings, whatever their particular culture. However, as MacIntyre (1998: 95, 96) warns, 'this kind of argument might be quite wrongly held to provide us with a kind of transcendental deduction of norms for all times and all places', for in seeking criteria to guide actual choices, 'the more I particularize my situation the more I ask for guidance for people who belong specifically to my time and place'. As we have already observed, it is working between the universal and the particular that morality is created, lived and understood. This is a central truth of ethics in a world of difference.

CHAPTER 3

LANDSCAPE, LOCATION AND PLACE: MORAL ORDER

All geographies are, in the last analysis, moral geographies. (Shapiro 1994: 499)

The concepts of landscape, location and place are fundamental to the way in which geographers describe and interpret the world. Landscape involves the visual appearance of an area, with its assemblage of natural features and human creations. Location is the position of something, expressed in grid co-ordinates, in relation to other things, or in such terms as near and far. Place denotes a site or portion of space in which objects and persons are often intimately related, as a setting for individual activity and social interaction. Each of these concepts captures different yet related facets of human life in its geographical particularity.

Interpretations of landscape, location and place involve normative judgements. Landscape tends to invite aesthetic reactions, an approving or disapproving gaze. Location(s) may be evaluated by efficiency criteria, like the least-cost location for a factory or point of minimum aggregate travel for a population. Place is often a locus of personal or group identity, which may be expressed as a sense of place highly valued among those who construct and experience it. All such interpretations have moral content, reflecting what is considered good (or bad) in human life in a geographical context. Indeed, the creation of landscapes, locations and places, as well as their interpretation, are intrinsically moral projects: imposing order or ugliness on the natural landscape, locating to an economic or social purpose, or imbuing particular places with value according to what they represent. This chapter explores some moral readings of the human geography revealed in landscape, location and place, as perhaps the most obvious exemplars of moral geographies.

MORAL READINGS OF LANDSCAPE, LOCATION AND PLACE

[L]andscapes have a moral dimension, they speak to notions of how the world *should* be, or more accurately how it should *appear* to be. (Cosgrove 1989: 104)

In October 1997 the Warsaw City Council rejected a plan by a Belgian developer to rebuild the Saski Palace (destroyed by German invaders in the Second World War) on Piłudskiego Square in the city centre. There was concern that redevelopment, involving a bank, would lead to commercialisation of the square, which contains one of Poland's most significant monuments. Grzegorz Zawistowski, Vice-Leader of the City Council and one of the dissenters, was quoted as follows:

> This is the place where the Tomb of the Unknown Soldier is and where national celebrations take place. It would be strange if all this happened in front of a Belgian bank . . . We want foreign investment in Warsaw, but there are certain places that shouldn't be commercialized, like the Tomb . . . Because of the connection between foreign capital and profit-making, we think that there should be more respect for something so nationally sacred. It is hard to imagine the owner of the Eiffel Tower not being French. (Grodsky 1997: 5)

However, a spokesman for the local voivodship (administrative district), Ryszard Hoffman, claimed that the Council's rejection of the plan was not simply a matter of concern for sacred ground, but was politically motivated, reflecting a conservative turn in recent elections. He recognised that any private sector development would bring commercialisation, adding: 'If these will be sacred places, they will be places without an income'.

These interpretations of the meaning of a particular place involve complementary and competing moral discourses, played out in the distinctive geographical and historical context of Poland's changing economy and society. A discourse of nationalism is involved in opposition to foreign investment, with a sub-text inviting recollection of earlier invasions. This is linked to a supporting discourse of anti-commercialism, associated with residual anti-profit sentiments inherited from socialism and now construed by some as conservative invocations of past values. The parochial and cosmopolitan are in conflict. So are the sacred and profane. And there is a distinctive moral geography in the implication that some places might be respected as gaps on the surface of pecuniary land values and profit. All this and more can be read into the debate on the future of this one city square. How such conflicts are resolved, within the broader process of redevelopment of a city in transition from one form of society to another, will be revealed in the extent to which a built form driven by the profit imperative of capitalism overlays the socialist (and presocialist) inheritance.

Monuments are particularly potent symbols of values. Whether recognising military triumphs (like Nelson's Column in Trafalgar Square), revered leaders (generals, monarchs), or honoured groups (such as the heroic workers in socialist city squares), their moral messages can be eloquent expressions of human ideals. They can range from the solid grandeur of the Basilica of the Sacred Heart in Paris (Harvey 1979) to the poignant spontaneity of the memorialisation of the site of Yitzhak Rabin's assassination in Tel Aviv (Azaryahu 1996). And as well as celebrating the good they can signify great evil, like that of the Holocaust (J. E. Young 1993).

When societies change, symbols of the old order may be toppled, like statues of Lenin in postsocialist Eastern Europe. Conflict may arise, as when the statue of South Africa's apartheid architect Dr Hendrik Verwoerd was removed from its plinth in Bloemfontein against the wishes of some Afrikaners. Place and street names may be contested, as those of communist revolutionaries are replaced by earlier princes or saints, who had themselves been erased under a former socialist regime. Thus St Petersburg, which became Leningrad, is now St Petersburg again. The use of ancient biblical or Talmudic references by the State of Israel to rename Arab villages in Hebrew is an important part of the struggle over land, as is the preservation of the original name in everyday Palestinian usage (Rubinstein 1991: 10; Cohen and Kliot 1992). All this is part of the exercise of political power, in pursuit of (or resistance to) some moral purpose associated with a particular way of life.

The interpretation of landscape is a long-established geographical practice. But since the middle of the 1980s the approach has broadened, with landscape being taken as a text to be read much as a written document or painting might be (e.g. Cosgrove and Daniels 1988). Landscapes have come to be seen very much as reflections of economic, political and social systems, with their prevailing ideologies or 'moral order' symbolised at the scale of settlement design as well as of individual structures. Thus, for example, 'liberal' and 'neo-conservative' landscapes may be identified in the city of Vancouver (Ley 1987). New cultural forms challenge the past in the eclecticism of postmodern architecture, just as the functional simplicity of modern design broke with the decorative extravagances of the Victorians. The collage of contemporary landscape might be seen as a representation of different changing and competing discourses, each deploying aesthetics to convey something of their moral content and strength.

Explicit attention to the moral significance of landscape, its creation and interpretation, is exemplified by Yi-Fu Tuan. He explains that the moral and aesthetic were inextricably linked in premodern times, as in China where the imperial park of the early dynasties was 'a world of natural and supernatural spirits' (Tuan 1993: 215). In the Western Hebraic-Christian tradition, natural landscapes were often portrayed as pure and free: moral qualities accrued to them, with God revealed in nature. The contrived association of the beautiful and the good could also serve more earthly purposes, for example in the 'aesthetic-moral state' of sixteenth-century Venice, with the pomp of sight and sound used to reinforce the power of rulers: 'the arts, including ceremony and architecture, present and dramatize the good' (Tuan 1993: 193). He also refers to the 'moral standing' of the rural landscapes which emerged in Europe in the seventeenth and eighteenth centuries, their displays of the megalomania, discipline and power of the landed aristocracy operating in a similar way to architecture in the city visibly confirming the status of the new capitalist magnates (Tuan 1989: 85–98).

The association of nature with particular values can promote political ideology in other ways. For example, in the United States the process of westward settlement gave the frontier a moral tone: 'When fused with the ideals of simple manners and love of liberty, it turns into a sort of geographic-aesthetic-moral icon' (Tuan 1993: 205). The city may be seen as the embodiment of social disorganisation, with the contemporary urban landscape oppressive: 'Rather than symbolizing a more open and democratic society, glass towers, for pedestrians pounding the sidewalks, have come to stand for exclusivity, indifference to public needs, technological fantasy, corporate privilege, arrogance, and power' (Tuan 1993: 150–1).

The geographer is not alone in finding links between the built form of the environment and prevailing power relations. For example, Michel Foucault has argued that architecture became political at the end of the eighteenth century, promoting the aims and techniques of governments:

> One begins to see a form of political literature that addresses what the order of a society should be, what a city should be, given the requirements of the maintenance of order; given that we should avoid epidemics, avoid revolts, permit a decent and moral family life, and so on. (quoted in Rainbow 1991: 239)

These objectives raise questions of the spatial organisation of the city, the construction of its collective infrastructure, the design of workers' housing and so on, in pursuit of social control.

Foucault saw less (if any) scope for spatial structures of liberation. However, some model communities which were built to promote collaborative economic activity, cooperative social life, or religious observance might be regarded as potentially liberating. Examples include some American experiments in collective living (e.g. by the Amish), and the Israeli *kibbutz*. The separate single-family home of modern, Western suburbia might be liberating, to those disliking collectivism. There is even the possibility of therapeutic landscapes (W. M. Gensler 1992).

The term 'moral geography' has been adopted in recent years to cover a variety of moral reading of geographical situations (D. M. Smith 1998a:

14–18). Some writers prefer 'moral landscape', 'moral location', 'moral terrain' or 'moral topography'. The earliest usage of 'moral' in this context appears to be by Peter Jackson and Susan Smith (Jackson 1984; Jackson and Smith 1984). Their examinations of the 'moral order' of the city, as revealed by the Chicago school of urban sociology in the 1920s and 1930s, focused on the work of Robert Park. He suggested that the city could be divided into natural areas, each occupied by their natural group, with every type of individual finding somewhere a 'moral climate' in which to flourish. Park and his students drew attention to those little worlds that departed most from the conventions of bourgeois society (the slums, ghettos, immigrant colonies, bohemias and hobohemias), finding a moral order in parts of the city conventionally thought of as among the most disorganised. Park concluded: 'Every natural area has, or tends to have, *its* own peculiar traditions, customs, conventions, standards of decency and propriety' (quoted in P. Jackson 1984: 175); the process of segregation established 'moral distances' which made the city of Chicago 'a mosaic of little worlds which touch but do not interpenetrate' (quoted in P. Jackson and S. J. Smith 1984: 66).

The notion of moral order has been extended to the Victorian city in Britain. The construction of subsidised housing for the 'deserving poor' is seen as an attempt to impose a moral order, similar to early experiments in urban planning such as company towns (P. Jackson 1989: 82; see below). Geographical space was deeply implicated in the struggle between the bourgeoisie and the popular classes: 'Moral order in the Victorian city was underpinned by its social geography (the segregation of classes and the separation of home and work, with its implicit gender division of labour)' (P. Jackson 1989: 100). Felix Driver (1988: 277) refers to the perception of the problems of the labouring classes, including pauperism, crime, ill-health, drunkenness, delinquency and degeneracy, as part of a 'moral constellation' or wider syndrome of anti-social behaviour. Moral miasmas associated with particular localities were viewed by the Victorians as contagious, like physical diseases, calling for a scientific response: 'The project of social science rested upon the identification of moral regimes and their relationship with environmental conditions. The mapping of the moral geography of the city, in particular, provided a basis for social intervention' (Driver 1988: 279)

There are links here with the work of David Sibley (see Chapter 6), who suggests that nineteenth-century schemes to reshape the city could be seen as 'a process of purification, designed to exclude groups variously identified as polluting – the poor in general, the residual working class, racial minorities, prostitutes, and so on' (Sibley 1995: 57). Faith in the possibility of amelioration through environmental reform was reflected not only in the building of model housing but also in the provision of such facilities as reformatories, hospitals, nurseries and public parks.

Specific case studies include documentary evidence of moral discourses underpinning two facility location decisions, or 'moral locations', in nineteenth-century Portsmouth. One is the relocation of the Soldiers' Institute, so that 'the men's recreational geographies could be shaped and their paths bent towards morality and away from immorality', which involved knowledge of the 'moral geography' of the city in terms of the 'centres of vice and disease' to be avoided (Ogborn and Philo 1994: 223, 224). The other is the location of a naval 'lunatic' hospital in a place which could direct the gaze of those within towards objects familiar to sailors: an exercise in 'moral geography' designed to divert the attention of those whose minds had become disturbed from 'immoral' facilities and neighbourhoods beyond Navy regulation (Ogborn and Philo 1994: 228).

Another case is the location of nineteenth-century reformatories (Ploszajska 1994). Cities were seen as sources of ills which could be ameliorated in rural settings, so reformatories came to combine a particular physical environment with the supposed virtue of hard labour. Thus a boys reformatory, the Philanthropic Society Farm School in Redhill, was located where it could be seen from railways (thus acting as a prominent deterrent), in countryside well out of London, with a regime of agricultural labour, spatially separated houses to simulate a family environment, and an imposing chapel to endorse religious habits. An institution for girls, Red Lodge Reformatory, Bristol, was removed from city life but suburban rather than rural so as to prepare girls for domestic life in their own families or in the service of those of others; again, the design of the facility was supposed to create a family atmosphere.

A variation on the link between environment and moral geographies is provided by the notions of the 'moral discourse of climate' and 'climate's moral economy' in the work of David Livingstone (1991, 1992). He explains:

> [D]iscussions of climatic matters by geographers throughout the nineteenth century and well into the twentieth century were profoundly implicated in the imperial drama and were frequently cast in the diagnostic language of ethnic judgement . . . the idioms of political and moralistic evaluation were simply part and parcel of the grammar of climatology . . . scientific claims were constituted by, and then made to bear the weight of, moralistic appraisals of both people and places.
>
> (Livingstone 1992: 221)

Thus, in the prevailing spirit of environmental determinism, native inhabitants of tropical lands could be described as in a low state of morality: sensual, unseemly, lazy and so on. And such behaviour as excessive use of alcohol, association with 'inferior' races and sexual indulgence characterised the 'moral topography of the white tropical experience' (Livingstone 1992: 236). Advice on how to survive in the tropics was couched in 'the moralistic language of prudence, abstemiousness, circumspection, and hygienic discipline. Here again was a moral economy of climate – not in this case that climate *conditioned* standards of behaviour, but rather *required* them' (Livingstone 1992: 240).

While much of the exploration of moral geographies, landscapes, locations and so on adopts a historical perspective, there are applications to the contemporary city. For example, cooperative housing in Vancouver, in the context of a 'moral landscape', has been described as 'capable of supporting humane, indeed moral, public values' (Ley 1993: 130), especially communal identity rather than individualism, yet it is personalised – in contrast to modern uniformity. The cooperative housing movement 'has sought to dislodge the alienating ethos of public housing and elevate an empowering vision of social housing' (Ley 1993: 145).

Other subjects for investigations of geographical expressions of morality include gender roles. A study of local childcare cultures with special reference to pre-school education in Sheffield has used interviews to identify 'a moral geography of mothering . . . a localised discourse concerned with what is considered right and wrong in the raising of children' (Holloway 1998: 31). In the affluent district of Hallam the 'good mother' knows a great deal about the range of available services and ensures that her children receive the best care possible, while in less prosperous Southey Green the 'good mother' sees that her child's name goes on the list for a popular local authority nursery. 'The moral geographies of local childcare cultures are important both in defining mothers as a social group and in influencing the meaning and experience of motherhood for individual women' (Holloway 1998: 47). Similarly, suburban differences around moral codes of mothering and participation in paid work have been observed (Dowling 1998).

Another topic which invites moral interpretation is sex. Recent work includes a study of 'red-light' districts in Birmingham, in which the terms 'immoral geographies' and 'immoral space' are used in the interpretation of conflict over the presence of prostitutes on the streets of Balsall Heath where local Muslim men protested at what they saw as a threat to their religious values (Hubbard 1998). This and similar work on prostitution (Hubbard 1997), on gay men and lesbians (Valentine 1993; Binnie and Valentine 1999), and also on children (Valentine 1996), raises the general question of the 'right' place for certain kinds of persons and practices (to which we return in Chapter 6).

Lest it be thought that this kind of perspective is confined to urban geography, reference can be made to studies invoking the contested 'moral terrain' of ancient forests and 'moral landscape' of the Pacific Northwest of the United States, by James Proctor (1995, 1999). He is concerned with 'how ideology transforms habitat into a moral landscape, a geographical embodiment of the good' (Proctor 1999: 193). He examines a debate over whether the spotted owl should receive special protection under the Endangered Species Act, which became a political struggle over whose moral landscape was to prevail: that of the environmentalists committed to preservation of the ancient forests symbolised by the spotted owl, or that of the timber industry and its dependent communities favouring resource exploitation.

Few writers on moral geography and similar concepts give much direct attention to the meaning of 'moral'. An exception is David Matless (1994), who is concerned with moral conduct in the explicit sense of different, often conflicting ways of 'being-in-the-world', and reactions to them. He suggests that the moral geographies of the Norfolk Broadlands work around Michel Foucault's three senses of morality: as moral codes, as the exercise of behaviour in transgression or obedience of a code, and as the way in which individuals act for themselves as ethical subjects in relation to elements of a code. Various texts from the late-nineteenth to the mid-twentieth century reveal conflict over the use of the Broads, in the context of boating (sailing portrayed as more commendable than motor cruising, the former encouraging positive attributes of individualism while the latter was more a noisy mass activity) and nature (the tension between shooting birds and watching them). 'The culture of nature, as much as the vulgar boating . . . is a matter of moral geography, of how people might or might not form themselves environmentally . . . social and aesthetic questions mingle in moral geography' (Matless 1994: 144, 146). Elsewhere, he shows how leisure pursuits in the English countryside have been presented in terms of the development of good citizenship (Matless 1997): another aspect of environmentalism as moral improvement.

Links with literature in moral or political philosophy are rare in these kind of moral readings. An exception is Gordon Clark's critique of some aspects of the formulation of justice elaborated by John Rawls (1971; see Chapter 7). Rawls's so-called original position, under which people decide on just institutions behind a veil of ignorance as to their actual position in society, is described as 'an imaginary moral landscape . . . a particular geography of morality . . . a very moral place . . . the ultimate color-blind, non-sexist, non-racial community . . . a utopia with a moral order we can only dream about' (Clark 1986: 147, 152). As other critics have recognised, but without the allusion to place, all this is relevant to the plausibility of the social contract which Rawls deduced.

No references to moral geographies, landscapes or locations have been found in the literature of moral philosophy. However, they have been appearing in some other fields (e.g. Shapiro 1994; see Chapter 6). Such applications tend to use these concepts unreflectively and in isolation from the geographical literature, an indication of failure to achieve interdisciplinary integration of work with strong thematic links.

A number of distinct but related themes may be found in studies of moral geographies and the like. The most obvious is the identification of moral differences: spatial variations in values and practices, from the 'mosaic' of Chicago to mothering in Sheffield. Another is the role of environment in moulding morality, reflecting the discipline's traditional preoccupation with behavioural responses to both nature and built form. Above all is a focus on power, and especially on interpretations suggested by Foucault: 'moral geographies operate not only through a dominative power of control and exclusion but also through performative powers of spatial practice' (Matless 1995: 396). Such studies might be regarded as exercises in 'thick' descriptive ethics (Proctor 1998a: 13), in which empirical observation fuses with contextual interpretation. They add substance to the abstractions of moral philosophy, helping to reveal different ways in which the right and the good is understood and acted upon, by real people in actual situations. Moral readings can also bring a new dimension to the interpretation of landscapes, locations and places. It is to a more extensive demonstration, in a particular kind of city, that we now turn.

THE MORAL GEOGRAPHY OF THE INDUSTRIAL CITY

It is hard to think of anything in human history with a geographical impact comparable to that of the Industrial Revolution. The industrialisation of Western Europe precipitated rapid urbanisation, while the shift from small-scale manufacturing to the more concentrated factory system introduced new elements into the landscape. But the multi-storey mills, mine head-gear, metals furnaces and so on were more than novel spectacles: they represented a new moral order, grounded in new sources of wealth and power, and in new social relations. The rows of cottages built for workers, which in some places consolidated as company towns, spoke of the reconstitution of labour's former feudal ties to employers. The prevailing political philosophy of *laissez-faire*, with its foundations in utilitarianism and free-market economics, let rip new forces under which both humankind and nature were obliged to yield to the profit imperative of capitalism. The landscape came to reflect a hard, material functionality.

The company town was a particularly potent expression of the values of industrial capitalism. The erection of workers' houses was often a necessity, arising from the remote and sparsely populated sites of early factories dependent on water power or localised materials. Such was the case with the settlements which grew up around the cotton mills established by Richard Arkwright and Jedediah Strutt in the latter part of the eighteenth century at Cromford, Belper and Milford in Derbyshire's Derwent Valley (D. M. Smith 1965: 67–72). Their primary purpose was to create a stable, content and therefore efficient labour force, and they soon began to take on the range of functions expected of conventional towns. Factory discipline was sensitive to the need to retain rather than repel reliable workers, whose health, religious observance and education were catered for. And the fact that some of the first housing is still occupied today indicates its quality. By the standards of their times the Arkwrights and Strutts were good employers, who 'felt responsibility as heads of their communities' (Fitton and Wadsworth 1958: 225–6). It was Richard Arkwright, in partnership with Robert Dale, who introduced the mill town to Scotland, creating what came to be portrayed as a moral exemplar in this far from benevolent age: New Lanark, under the management of Robert Owen.

Such precedents, from the earliest days of industrial capitalism, came to their most elaborate fruition in model company towns of the second half of the nineteenth century. Early experiments in town planning, such as Saltaire (1853), Bournville (1893) and Port Sunlight (1888), were 'attempts to translate the paternalistic impulse of an earlier generation into a form that was more suitable to the Victorian city' (P. Jackson 1989: 82). They revealed their own distinctive moral geographies. For example Sir Titus Salt provided shops, alms-houses, a hospital, a church, and other public buildings, but no pubs as he preached against the evils of alcohol (Minchinton 1984: 131). Religion was a strong motivating force in the Quaker Cadbury family's garden factory village of Bournville. Edward Cadbury shared with Port Sunlight's founder William Lever the belief that business efficiency and the welfare of employees were different aspects of the same problem.

Similar sentiments and strategies appeared in the United States. At Pullman, Illinois (founded in 1880), the rail-car magnate George Pullman built houses, tenements, shops, offices, schools, stables, playgrounds, a market, a hotel, a library, a theatre and a church: 'a model town, a planned community' (Walzer 1983: 294). George McMurtry's model industrial town of Vandergrift, Pennsylvania, was described by a resident in 1901 as 'moral and modern'; the goal was to create a settlement that would house loyal non-union workers, involving the ideal of home ownership, and of environmentalism expressed in expensive landscaping (Mosher 1995: 101).

The remainder of this section develops a case study, further to explore some of the issues raised by the moral geography of the mill town. The city of Łódź in Poland, with a population of about 850 000, is second in size to the national capital of Warsaw eighty miles to the east. Łódź grew up on the prowess of a cotton textile industry, which earned it the accolade of the 'Polish Manchester'. Here the company town developed in a distinctive setting, giving rise to a landscape eloquently reflecting both the power and the paternalism of capital.

At the beginning of the nineteenth century Łódź was a small agricultural village (see Figure 3.1). A new town was laid out on adjoining

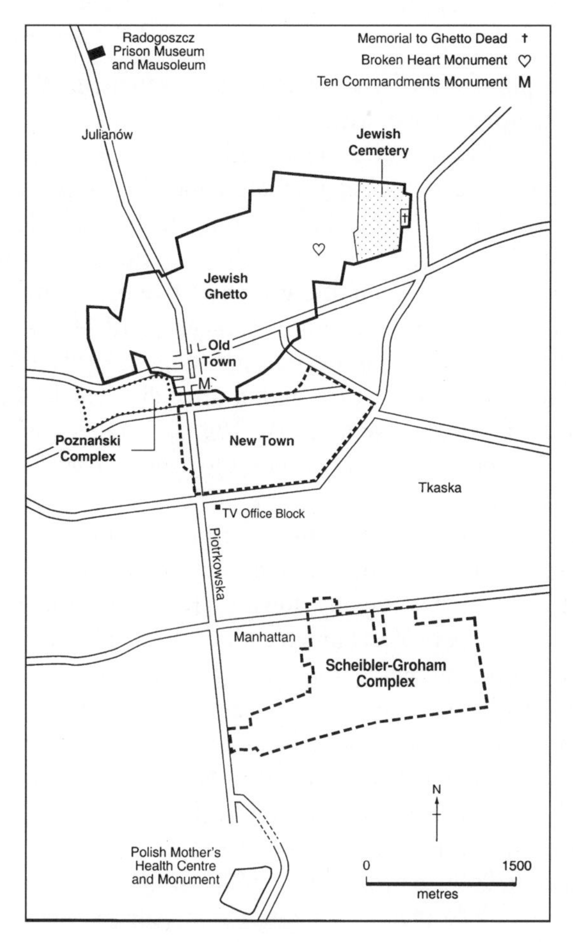

Figure 3.1 *Some features of the landscape of the city of Łódź. (Source: boundary of Jewish Ghetto, 1941–44, from Adelson 1996: 17)*

land in the 1820s, at the instigation of the Polish government, in the hope of stimulating the manufacture of textiles. Subsequent success was beyond all expectations, and the population increased from 10 000 in 1840 to 320 000 at the turn of the century. By 1939 the city of Łódź had 680 000 inhabitants, and accounted for 37 per cent of employment in the Polish textile industry (for further details, see: Puś 1991; Liszewski 1997).

The landscape which emerged had no precedent in this part of Europe:

> Within this jumbled mass were all the contrasts which we have come to accept as typical of towns created in the nineteenth century to serve mill industry . . . The dark, insanitary and overcrowded housing of the vast majority of people was punctuated on the one hand by flamboyant 'palaces' belonging to the mill owners who, in Łódź, seemed to prefer their mansions to be near their mills rather than outside the town, and, on the other, by the few surviving cottages of wood and thatch which had been built by the first textile masters in the 1820s.
>
> (Dawson 1979: 379)

Most of the population lived in multi-storey tenement blocks with a central court, interspersed with factories generating smoke, fumes and noise. The city of over half a million inhabitants had no waterworks or sewage system before 1914. Yet gas illumination had appeared on selected streets at the end of the 1860s, the Bell company had installed a telephone network in 1883, and the first electric lighting in Poland was introduced in some factories and villas soon afterwards. In five of the fourteen districts into which Łódź was divided, more than one-third of dwellings had an average occupancy of over four persons per room in 1931; only 10 per cent of all dwellings had a water closet, and in three districts the figure was less than 1 per cent (Tomaszewski 1991: 193–4). Andrzej Wajda's film of Władysław Reymont's novel *Ziemia Obiecana* (*The Promised Land*) captures something of the city's era of predatory capitalism, which produced large fortunes and provoked celebrated strikes.

The most distinctive feature of the development of Łódź comprised enclaves of individual company towns, created by the major factory owners in the latter part of the nineteenth century. The largest was that of the Scheibler and Grohman Corporation (Figure 3.1): the complex known as Księży Młyn (Priest's Mill), which claimed to be second to none of its kind in Europe. This incorporated factory buildings, workers' dwellings, a hospital, school, kindergarten and other facilities including its own fire station, and also a farm (Figure 3.2). The first housing had been built in 1860; by the beginning of the twentieth century, when the complex employed about 9000 persons, over 2600 families were accommodated there. The estate also included mansions or 'palaces' for the mill-owner and his family. Among them is a neo-Renaissance villa of great architectural distinction, built in 1895–7 for the founder Karol Scheibler's

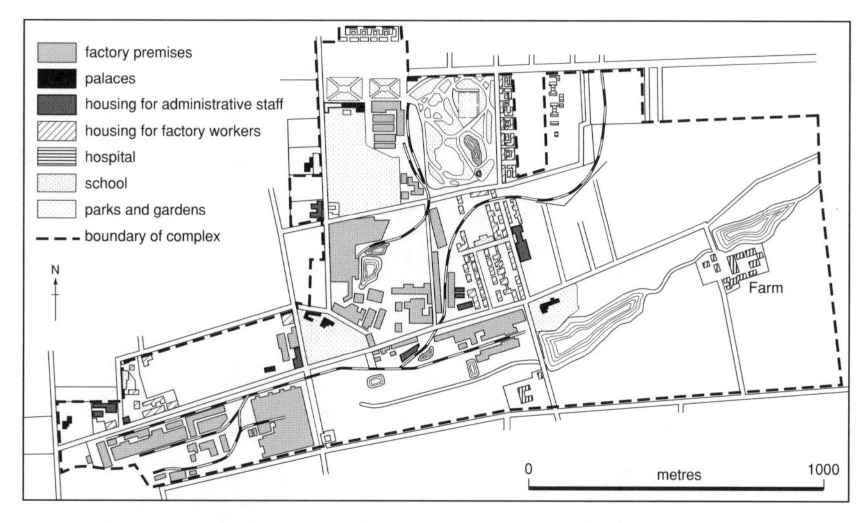

Figure 3.2 *Layout of the Scheibler-Grohman factory complex of Księży Młyn in Łódź. (Source: Liszewski and Young 1997: 59)*

son-in-law and factory manager Edward Herbst, situated in a sumptuously landscaped park.

The moral order represented at Księży Młyn had a number of dimensions. The massive and imposing redbrick facade of the mills dominated the neat rows of workers' dwellings with out-houses at the rear (Figure 3.3). Symmetry was imposed by an axis extending along a street from the main entrance. Superior houses for technical and administrative workers occupied separate sections (Figure 3.2). The palaces, of varying styles and status, generally fronted onto a street, the better to display their owner's wealth and taste expressed in elaborate facades of the latest architectural styles, while adjoining gardens provided privacy almost comparable with that of a country estate. A subtle blend of class coexistence and separation is suggested.

The second of the great factory complexes was that of the Jewish entrepreneur Izrael Poznański (Figure 3.1). His business expanded rapidly from the late 1870s, to employ about 7000 workers by the beginning of the First World War. Enormous factories occupied most of a site more compact than that of Scheibler–Grohman. Apartment complexes were built for about 1000 workers' families, but predominantly tenements with gloomy interior courts: inferior to the housing at Księży Młyn, and lacking social facilities except for a concert hall, hospital and church. Poznański's massive palace, fronting conspicuously onto two streets, was started in 1877 and displays exuberant neo-Baroque splendour to the external public gaze while offering a more restrained exterior to the private garden. The

Figure 3.3 *The Scheibler-Grohman complex with mills and workers' houses, Lódź, at the end of the nineteenth century.*
(Source: album published by the Scheibler and Grohman Company, 1921)

abrupt change along one of the streets (Ogrodowa), from the stucco palace facade to that of the redbrick mills with monumental entrance, fittingly captures the aesthetically uneasy but functionally imperative association of the ostentatious outcome of wealth with its more mundane origin (Figure 3.4). The landscape link with labour, living in tenements across the street, was less direct than at Księży Młyn where housing was integral to the complex.

Figure 3.4 *Izrael Poznański's palace and adjoining cotton mills on Ogrodowa Street, Łódź. (Source: Author's collection)*

Another feature of the landscape of capitalist industrial and commercial development is extravagant architecture of residential buildings of sub-palatial status (Popławska and Muthesis 1986). Most of these are along the main street of Piotrkowska (Figure 3.5), and those which join or run parallel to it: tenement blocks, up to seven storeys, with narrow elevations to the street and courtyards behind. The facades were often given elaborate treatment, some neo-Renaissance, some neo-Baroque, but most distinctively in a modern style known as 'Łódź secession'. But, as with the mills, their architecture carries an international flavour, owing nothing to local culture (Riley 1998: 101). Just as Łódź was *de facto* an international city, with its substantial German and Jewish populations added to the Poles, so it entered the present century with a landscape embracing the cosmopolitan style of the modern age.

The landscape also reflected the consolidating class structure of capitalism. The most successful manufacturers and bourgeoisie were able

Figure 3.5 *Late nineteenth-century facades on Piotrkowska Street, Łódź. (Source: Author's collection)*

to display their status on street frontages, for those without the privilege of access to their sophisticated interior design, furnishings and art collections. The association of new wealth with aesthetic refinement reflected a bourgeoisie without a historical past patronising the arts in imitation of the aristocracy (Szram 1984: 283, 286). Even the tenements provided scope for differentiation, and separation. The wealthier part of the population lived in the front flats, larger than the others, better appointed and with better facilities, while extensions behind were filled with poorer persons, often villagers who had come to the city seeking work. Thus there was social stratification at this most local scale.

The link between industrial development, technological innovation and the arts in Lódź created a landscape in which mills may be admired for stylistic embellishment as well as enormity of scale, where a model factory complex could be compared favourably with any of its contemporaries elsewhere, and where the homes of the owning class justify the customary local description of palaces. The work of architects of the distinction of Hilary Majewski is found in public buildings and former private homes along Piotrkowska and other main streets, as well as in the most magnificent mansions. And as in life, so in death, the major factory masters are memorialised in tombs of competing grandeur displaying the high aesthetic fashions of the time, most prominently in the Scheiblers's chapel in the Evangelical cemetery and in the mausoleum of Izrael Poznański, dominating part of the Jewish cemetery (Figure 3.6).

Figure 3.6 *Izrael Poznański's mausoleum in the Jewish cemetery in Lódź. (Source: Author's collection)*

Łódź entered the era of socialism following the Second World War with its urban fabric largely intact. The new political and social order was therefore unable to impose itself on, and through, the landscape with the freedom provided by the wholesale destruction of Warsaw, for example, or by the rapid construction of major new industrial projects like Nova Huta near Kraków. The workers' tenements remained, as did the factories, shops and offices. Moreover, 'the planners' fixation with industrialisation as the panacea for economic growth, coupled with the low priority accorded to housing, ensured a minimum of change during the following four decades' (Riley 1997: 456). The mills kept their original use, and machinery, for the most part, following the state's decision to maintain the city's specialisation in textile manufacturing. Only the mill-owners' homes, some apartments of the bourgeoisie, and most of the banks experienced a change of use. But the main features of the pattern of land use laid down under capitalism remained.

What socialism added to the inner-city landscape was modest, and its moral message was more restrained than under capitalism. Some high-rise office blocks were built in the 1970s, as a southern extension of the central business district, along with new department stores and residential blocks substantially higher than Łódź was accustomed to, skirted by a four-lane highway out of scale with the city's trickle of traffic. These concrete-and-glass boxes of the times, in a district which became known as Manhattan, did less to evoke socialism than did the subsequent years of negligible maintenance. Of greater symbolic significance was a new office complex containing a fifteen-storey tower block more centrally located, in a style of Stalinist brutalism reminiscent of Warsaw's Place of Culture and Science (a 'gift' to Poland from the Soviet Union). But the Łódź edifice (occupied by Polish Television) is not on the same dominating scale as its Warsaw counterpart, which came to symbolise not the friendship but the threat of the neighbour to the east.

The socialist period had more impact on the landscape of outer parts of the city, as it expanded with further population growth. New industrial districts were built in the 1970s, to facilitate construction of those huge plants which symbolised the power of industry under communism. New housing estates accompanied them, following the form prescribed by socialist urban planning in the Soviet Union, with neighbourhood units (or 'microregions') incorporating social services and cultural facilities along with their apartment blocks. Grandest of all is a hospital complex known as the Polish Mother's Monument, completed in 1988 (Figure 3.7), the significance of which is to be found as much in the city's reputation for maternal health problems associated with harsh working conditions as in the status of mothers under socialism. The outer areas represented the

Figure 3.7 *The Polish Mother's Monument hospital complex in Łódź. (Source: Author's collection)*

typical socialist city (French and Hamilton 1979; Smith 1989), with its local employment base, common housing standards and state-provided means of collective consumption: a moral landscape inspired by egalitarian values and an ethic (if not a practice) of distribution according to need, in contrast to the class inequalities of capitalism.

It is common to criticise the socialist city, as colourless, drab and uniform. One disaffected observer even suggests that the vitality of Communist faith in Poland can be traced by the standards of the regime's buildings: 'sub-Mussolini heroic style with stone facades in the 1950s, sober brick in the 1960s, shoddy concrete blocks in the 1970s, and nothing at all in the 1980s' (Sikorski 1997: 20). Like mass public housing in such countries as Britain and the United States, socialist urban structure was to some extent a reflection of the broader contemporary demise of modern idealism. Thus: 'After 1945, modernist architecture was largely deprived of its social vision and became increasingly an architecture of power and representation. Rather than standing as harbingers and promises of the new life, modernist housing projects became symbols of alienation and dehumanization' (Huyssen 1990: 239). However, if the comparison is with what went before, in cities like Łódź, then, for all their technical defects, the suburbs of the socialist era were a great advance on the housing and services inherited from the nineteenth and early twentieth centuries.

Two possible exceptions should be noted, in addition to the superior accommodation of the presocialist élite which was expropriated, subdivided and redistributed after 1945. One is the scattered neighbourhoods of single-family villas with gardens, mostly built in the 1930s. They are represented in Łódź by the district of Julianów and in Tkaska and adjoining streets (Figure 3.1). A survey of residential preferences in Łódź revealed Julianów as an idealised representation of high living standards, incorporating the dream home of many respondents (Kaczmarek 1995: 26).

The other possible exception is the industrial-residential complexes of the major capitalist mill-owners, exemplified by Scheibler's Księży Młyn, the wider moral significance of which may now be considered. Bolesław Domański has made comparisons between the role of the capitalist and socialist enterprise, with respect to social provisions for their workers:

> [I]t was typical of the [socialist] industrial enterprise in Central and Eastern Europe to provide a wide range of non-wage benefits through company facilities and services. While this contrasts with practices in modern Western corporations, it bears a striking resemblance to the practice in some early capitalist firms, whose broad involvement in the local community has often been referred to as 'paternalism'. (Domański 1992: 353)

This kind of relationship under capitalism might be interpreted as a particularly effective means of social control:

> Paternalism facilitated the preservation of the superior position of the owner/manager; moreover, it contributed to the legitimation of the employer's power and of unequal social relations. This occurred as a result of the dependence of workers upon elite definitions of their social situation, including their rights and obligations. The rhetoric of mutual personal duties masked the asymmetry of exchange. (Domański 1997: 59)

The Scheibler complex was surrounded by an ornamental cast-iron fence, with a porter at the gate. Old inhabitants recall the lawns and flowers, with people sitting and chatting on benches, yet: 'This estate, which was self-sufficient economically, in a way trapped workers, who became totally dependent on the factory owner' (Jabłoński and Jabłoński 1995: 13).

The significance, and legitimacy, of the spatial extension of ownership and control beyond the factory was famously challenged at Pullman in the United States (above, p. 54). George Pullman saw the building of factories with a town around them as just another business venture, and ruled the settlement with autocratic power. But, benevolent or otherwise, he had not been chosen by those he governed. Workers went on strike in 1894, leading the Illinois Supreme Court to order the Pullman Company to divest itself of all property not used for manufacturing, on the grounds that the ownership of a town was incompatible with democratic institutions. 'What democracy requires is that property should have no political currency, that it shouldn't convert into anything like sovereignty, authoritative command, sustained control over men and women' (Walzer 1983: 298). In short, there should be limits to the spatial scope of the power of capital. However, the profit imperative drives capital to transcend such limits, even at the expense of the democratic values it often claims to uphold.

There may have been similar tendencies for the socialist enterprise to extend the spatial scope of its power, through 'company enclaves' within the city:

> [A] considerable part of paternalistic provision took place outside the limits of the plant. Industrial child care, sports and medical facilities, and entire apartment complexes located in various urban quarters or in downtown remained extensions of the company beyond the control of the local authorities, not to mention the community itself, and usually benefited the general economy/community little. Apart from their functional significance for the life chances of certain groups, these extensions became symbols of company power. (Domański 1997: 160)

This is an example of the concept of territoriality as a strategy of controlling people by controlling territory (Sack 1986), in which the power of the 'gatekeepers' who provided access to company benefits was crucial. At the extreme, the entire town could be a company enclave, as in some industrial settlements in the Soviet Union, which could help to absolve enterprises from responsibility for negative externalities such as pollution. However:

> [W]hat was sought in most paternalistic action was control over the local labor market, and spatial patterns were a vital mediating factor in the mechanisms of this control. The company's power underlay gatekeeping practices, which, in turn, through particular spatial and social configurations of local phenomena, reproduced this power. (Domański 1997: 174)

Thus, the moral geography of the industrial city under both capitalism and socialism reveals strategies to extend the spatial scope of producer power. While under capitalism the dynamics of this process would have been mediated by the market, and occasionally challenged by the workers, it is not clear how far local authorities or central state agencies might have constrained the power of individual enterprises under socialism. And the ultimate question of whether the overarching ideology of egalitarianism might have required enterprises to serve the people of cities like Łódź better than under paternal capitalism was consigned to historical speculation by the collapse of socialism itself.

The past decade has seen the emergence of a new society in Eastern Europe. The contours of its moral geography are by no means clear, but some elements in the landscape of Łódź are suggestive, and possibly surprising. In the city centre, functional areas capable of fairly clear definition existed under socialism, even in the absence of rent; far from promoting spatial specialisation, market forces have subsequently blurred earlier land-use patterns (Riley 1997: 454). Manufacturing has almost entirely disappeared from the central area, with the collapse of the textile industry, to be replaced by wholesale trading, retailing and offices.

The most obvious changes have been on the main commercial artery, Piotrkowska, always the place where prevailing values would be conspicuously on display. The street has been repaved and pedestrianised, with fast food and drink outlets spilling onto the pavement. McDonald's symbolises the arrival of American gastronomic refinement. The uninviting state shops have given way to outlets for prestige products, many bearing well-known Western brand names. Some of the more elegant facades have been refurbished. The once-gloomy courtyards behind reveal yet more new businesses: almost half of them in the legal profession (Riley 1997: 465). The pre-existing urban fabric, here and on other central streets, readily adapts to the proliferation of small private enterprises, with little new, purpose-built construction.

The ambience of Piotrkowska is increasingly that of a Western European shopping street. There is a sense of the kind of invasion to which opponents of foreign investment in Warsaw's Piłudskiego Square were so sensitive (see above, p. 46). The names on shops, signs and advertisements increasingly feature the English language. Surveys in the early 1990s showed local people disapproving of the way Piotrkowska was changing, just as some of St Petersburg's citizens felt alienated by similar changes on their main street, Nevskiy Prospekt (D. M. Smith 1994b: 621), but by the middle of the 1990s more people were approving – perhaps because of a different kind of population able to take advantage of the new quality retailing (Sylwia Kaczmarek: personal communication).

The most dramatic expression of the explosion of market forces is in open-air trading, with massive, largely informal markets on the edge of the city. Among those attracted to these sites are criminal gangs; increasing crime in Łódź is attributed to the growing disparity between rich and poor and to the high rate of unemployment (Wolaniuk 1997: 204–6). The market carries mixed blessings.

The landscape impact of housing privatisation is discreet, for the most part. Some new apartments are sprouting embellishments suggestive of affiliation with the postmodern. The most significant additions are in the number of single-family homes, and in the development of enclaves of luxury residences; people with the money may now realise their dream. As inequalities widen and socioeconomic stratification consolidates, this is reflected in increasingly differentiated housing stock, and in local environmental quality as superior services follow purchasing power.

An obvious visual expression of change has been the appearance of colour: in architecture, advertisements, shop-window displays and so on. To the superficial gaze, this more than anything else may differentiate the postsocialist city from that of the earlier era. But the observer more familiar with the nuances of social transition sees something else:

> One Western school of thought looks with some pleasure upon the 'regeneration' of the cities in eastern and central Europe as they emerge from their long sleep. They see the buskers, hoardings, street life and graffiti, private kiosks selling everything under the sun, shoe-shiners and the new uses to which foyers of administrative buildings are being put as some sort of triumph of the spirit . . . But with the colour have come the Hogarthian scenes of homeless people, beggars, refugees, child prostitutes, high levels of crime and hard drug use, gangland shoot outs and rampant corruption even among the organisers of charitable foundations. As the grey socialist city passes away, it leaves behind buildings with damaged stucco, fallen balconies and peeling paint, streets piled with uncollected garbage, once clean and pleasant parks and public spaces full of the detritus of the capitalist city. (Andrusz 1996: 29)

The new society has embraced the market with uncritical zeal. Among those things more conspicuously for sale than in the past is sex (there is a sign for a 'sex shop' on Piotrkowska). Mainly confined to the big hotels under communism, prostitution is now part of the new moral geography: on the city streets, the inter-city highways, and around border crossings where men from the other side can indulge particular preferences at bargain prices. It has been suggested that women get moulded by male violence into a role consistent with particular social requirements: 'every political change must re-enact the original violence against women required to sublimate them into a mirror image of their society' (Geyer-Ryan 1996: 3). Thus, the systematic rape of Bosnian Muslim women by Serbian men in former Yugoslavia was like conquering territory, given the prescription of female monogamy by Islamic law. Perhaps the visiting Western businessman renting a Polish woman for a while represents the ultimate penetration of market forces into the heart of a society which, under communism as much as Catholicism, idealised the image of the female as mother.

The people of Łódź may be more fortunate than those of many other postsocialist industrial cities, especially in the former Soviet Union, but the dangers are evident. While the minority *nouveau riche* may indulge their desires in fine homes, restaurants and fashion salons, for many others the old security of limited expectations and satisfaction of basic needs has been replaced by growing uncertainty and vulnerability. What postsocialism seems to offer is capitalism without paternalism.

As to the future of the great nineteenth-century mills, this is most likely to be as attractions for the visiting industrial archaeologist or for the heritage tourist, with the Scheibler and Poznański complexes perhaps extolling the long-lost virtues of paternal capitalism. In death as in life, moral mills. It is such surviving physical structures as these which can so selectively represent past moral geographies. The human struggles, and the suffering, are nowhere to be seen.

A MORAL GEOGRAPHY OF ABSENCE: THE ŁÓDŹ GHETTO

With the emergence of Łódź as an industrial town, large numbers of Jews began to move in. They were initially confined to part of the Old Town (Figure 3.1), but residential restrictions were removed in 1862, when they made up about a quarter of the total population. Subsequent Jewish immigration kept pace with the rapid growth of the city's population as a whole. Jewish businessmen, led by Izrael Poznański, had a major influence on the growth of the city (see *Polin: A Journal of Polish-Jewish Studies*, 6). By the criterion of native language Jews comprised 31.7 per cent of the city's population in 1931, compared with 59.0 per cent Poles and 8.9 per cent Germans (Tomaszewski 1991: 175–7); they lived mainly in the northern part of central Łódź, but were present in all districts. Jewish entrepreneurs of inter-war Łódź played an important part in the Polish economy. Jewish cultural life produced figures of distinction, the best known being the pianist Artur Rubinstein.

The city faced the Second World War with a Jewish population of 230 000: the largest such community in Europe except for Warsaw. In September 1939 the Germans annexed Łódź into the Greater Reich, and concentrated Jews in a run-down area to the north of the city centre. The Litzmannstadt Ghetto, carrying the city's new German name, was one of the first of its kind, and was officially sealed off in May 1940. To the city's original Jewish population was added 20 000 from elsewhere. Their fate has been recorded in contemporary accounts (Dobroszycki 1984; Adelson and Lapides 1989; Unger 1995), in the diary of a young woman who survived (Eichengreen 1994) and of a young man who died there (Adelson 1996). It is sufficient to record here that, with very few exceptions, those who avoided murder or starvation in the ghetto itself went to the death camps at Chełmno and Auschwitz. When the city was liberated in January 1945, less than a thousand Jews remained.

The single most remarkable feature of the moral geography of Łódź is, arguably, not the presence of those spectacular structures inherited from a great industrial past, but an absence: the absence of the Jews. How can we describe an absence? Not a case of nothing there, but of something gone. A moral geography sensitive to absence has to recognise that all that matters is not inscribed in the landscape and patterns of the present. Its representation has to come to terms not only with the death of those quarter of a million, and the survival of so few, but also with the absence of evidence or even recollection that most of them ever lived.

In Charles Powers's novel *In the Memory of the Forest*, set in a small Polish village typical of many whose Jewish community was erased in the Holocaust (see Chapter 4), one character says to another:

> Let me ask you, do you see any sign that they are missed? Do you see even a sign that they were once here? Any indication of who built the houses along the streets? Do you hear any of the old ones, even, remember the *challah* [sabbath bread] from Klemsztejn, the baker? The quiet on the streets on Friday afternoons? Do you hear them remember that there was once a man in the village who knew how to repair shoes? Or sew a coat? Do you see any marker for them? Any stone left where their dead rest? (Powers 1997: 431)

Even the cemetery in such a place is likely to have been desecrated, or used as a source of building material. In town after town, village after village, all that remains are a few upright stones bearing strange characters in an overgrown field, and a storehouse or cinema with decorated facade signalling that it was built for a higher purpose.

The Jewish cemetery in Łódź was spared. With more than 180 000 tombstones, it rivals that of Warsaw. They stand row upon row, some simple, some ornate, many tilted or broken, receding into the woodland which has recolonised the site. In the centre is the awesome mausoleum of Izrael Poznański, dominating the dead as he did the living. There are a few memorial tablets on the wall around the entrance. One reads: 'In loving memory of my dear sister Bluma . . . who was shot dead by a German murderer without any reason in the Ghetto Łódź Poland on 20 July 1940.' Outside the entrance, a small monument to the ghetto dead stands on a patio in need of repair (Figure 3.8). As in Warsaw, it is as if the extermination of the Jews had killed those in the cemetery a second time: 'it had killed the memories in which they might have lived on' (Todorov 1999: 4).

The Nazis erased virtually everything else in the city conspicuously associated with the Jewish people (Liszewski 1991). Three imposing synagogues were set on fire in 1939 and pulled down, though an elegant neo-classical structure built as a synagogue in courtyards behind apartment buildings on Piotrkowska still stands. The factories, palaces and other housing built by Jews remain, indistinguishable from those of the Gentile population. As for the ghetto itself, some old housing survives, but much of the area was subsequently covered by estates of the socialist period.

After the war survivors of the Holocaust gathered hundreds of tombstones together to use for a monument, but they could not agree on the final conception and left Łódź, their project abandoned (Young 1993:

Figure 3.8 *Differential memorialisation in Lódź: the Radogoszcz Museum and Mausoleum (left) incorporating ruins of a prison where 2000 Poles were killed, and the smaller and less conspicuous monument to more than 200 000 Jewish ghetto dead (right). (Source: Author's collection)*

194–6). The small monument built later on the eastern side of the cemetery, right on the ghetto boundary, is as isolated as it could possibly be (Figure 3.1). More conspicuous is the Broken Heart Monument to the west of the cemetery, on the site of a notorious Nazi prison camp for children. And far more imposing is the Radogoszcz Museum and Mausoleum to the north of the ghetto area, commemorating the martyrdom of citizens of Łódź, incorporating part of a Nazi prison for Poles set on fire, killing two thousand inmates, when the Germans left the city in January 1945 (Figure 3.8). It was 1995 before the Ten Commandments Monument in the form of a statue of Moses was erected, on the southern border of the ghetto, at the instigation of the Foundation to Commemorate the Presence of Jews in Polish History set up in 1991 (Figure 3.9). When I visited the monument in June 1999 its inscription had been defaced by a swastika.

To understand this selective memorialisation in the moral geography of Łódź, it is necessary to recognise that it has not been easy for Polish people to come to terms with the Holocaust. Some of this might have to do with

Figure 3.9 *Statue of Moses featured in the Ten Commandments (or 'Decalogue') Monument, on the site of the Łódź ghetto. That the monument was designed by a famous Polish sculptor (Gustaw Zemla), and stands on land provided by the City Council, is indicative of recent recognition of the Jewish past. The flats in the background typify the residential landscape of the socialist era, which replaced most of the ghetto. (Source: Author's collection)*

guilt, individual and collective, that more was not done to protect Jews. It is also related to the claim that very many Poles died in the war, but that it is the Jews who are remembered, and mourned, especially by the world outside Poland. For three decades, the Holocaust tended to be denied, hidden or forgotten. Communism gave the Polish people plenty of other things to worry about, including the erosion of their religiosity and hard-won sense of nationality, and there were few Jews left to tell their story. A documentary film (*Shtetl*, by Marian Marzynski) featuring the return of two survivors to their former home town of Brańsk, which had a majority Jewish population before the war, concludes with the revelation that a commemmoration stone, erected in 1993 on the five-hundredth anniversary of the town's foundation, carried no reference to the Jews. And they were not mentioned by the vice-mayor in his speech on the occasion, even though he studied Brańsk's Jewish history and had been responsible for the partial reconstruction of the Jewish cemetery (see Hoffman 1998). Such are local Polish sensitivites, still.

There were exceptions to this collective amnesia, like the erection of a monument amidst the rubble of Warsaw in 1948 on the fifth anniversary of the heroic ghetto uprising, and the public presentation of the Auschwitz concentration camp (Charlesworth 1994). But it was the late 1980s and early 1990s before acknowledgement and rediscovery of Poland's Jewish past began to build up (e.g. Kagan 1992): 'a reintegration of Poland's lost Jews into Polish national heritage' (J. E. Young 1993: 117). This was facilitated by political changes, and easier travel by Jews from elsewhere seeking their roots. Stephen Spielberg's film *Schindler's List* stimulated Holocaust tourism, which has seen the old Kraków ghetto of Kazimierz turn from shabby backwater to fashionable night-spot with fancy Jewish restaurants supplementing the ancient cemetery and synagogues (see below, p. 90 and Figure 4.1).

It is in such places as Łódź, Kraków and Warsaw that horrendous evil was somehow made possible. Eva Hoffman explains, in an account of the rediscovery of the Polish Jewish community of Brańsk (see Chapter 4):

> In relation to Jews in particular, the Nazi occupation created a world of monstrously inverted morality. It was a world in which the ordinary qualities of decency, responsibility towards others, concern, and compassion were criminalized, and in which rank brutality and sadism were normalized. We must imagine this: a pastoral town or village in which life looked, on the surface, almost normal, but which had actually been turned into a zone of legalized perversion – a zone in which the indigenous population, not very sophisticated or educated, was rewarded (albeit poorly) for selling the lives of its neighbors and killed for helping them. (Hoffman 1998: 241–2)

There was a widespread perception that Jews were beyond the Polish 'universe of obligation' (Fein 1979: 33). A Pole who took risks for a Jew was therefore stretching compassion beyond the bounds of responsibility (Hoffman 1998: 247), challenging the prevailing moral geography. But some did. We will return to the implications in subsequent chapters, as we seek further contextual understanding of the human propensity for evil, and for good.

CHAPTER 4

PROXIMITY: LOCALITY AND COMMUNITY

> If we should find ourselves unable, when the moment comes, to meet the stranger's gaze – and to be moved by it – then woe to him who is lost, who has wandered far from his people. (Todorov 1999: 296)

Summarising moral lessons of life in concentration camps, and of reactions to them, Tzvetan Todorov underlines the reluctance to do for strangers what we would gladly do for those we love. This includes the risks some persons were prepared to take to aid others, like those Jews trying to hide their identity, but recognisable by the look of sadness in their eyes. To be a stranger away from home, out of place, adds to the vulnerability of difference, to make special demands on human sympathy.

This chapter explores issues relating to proximity, in seeking to elucidate the significance of locality to how persons should treat others. Proximity means closeness or nearness. Locality usually refers to the site of something; here, it is taken to be a place of social relationships manifest in partiality, in the sense of a morality which favours close people over more distant (and different) others. Partiality was a convention of premodern society, strong enough to withstand the challenge of the Enlightenment insofar as much actual human practice is concerned. It is an aspect of localism, in the sense of favouring what is local, whether this be persons or prevailing values.

Locality is the customary context for the formation of communities. The term 'locale' (Giddens 1979) captures this sense of a setting for a special kind of social interaction. The geographer's long-standing interest in the notion of community, highlighted in a theme issue of *Environment and Planning A* (Silk 1999), leads to some central issues in moral and political philosophy, including the communitarian challenge to liberalism and the feminist-inspired ethic of care as a challenge impartiality. It also leads on to the question of how far the sympathy persons customarily bestow on their nearest and dearest may be extended, in the direction of universality, a fundamental geographical issue in ethics taken up in the next chapter. (Chapters 4 and 5 rework arguments originally set out in D. M. Smith 1998b and 1999a.)

LOCALITY AND PARTIALITY

The increasingly known and interdependent world, which today we take for granted, is a very recent phenomenon in human history. To favour people in close proximity was an understandable convention of the small-scale societies which prevailed, for the most part, until the modern era (as we saw in Chapter 2). Attitudes to strangers protected local group security: prudence born of experience tended to prevail over more altruistic sentiments. Those posing a threat were repulsed. Care for outsiders was confined to codes of hospitality, which could be quite specific as to the extent and duration of sustenance and shelter which hosts were obliged to provide before the visitor had to explain or move on.

Localised partiality was a feature of the ancient world. In Greece, obligations to one's city, fellow citizens, associates, friends and family were recognised, but 'the question was always likely to arise why one should honour obligations whose strength appeared to stand in inverse proportion to their distance from home' (Rowe 1991: 125). Aristotle insisted that to try to extend the bounds of familial love to everyone destroys family bonds themselves (Tronto 1987: 659). He believed that, while Greeks should not enslave each other, it was not morally reprehensible to enslave 'barbarians' whose language and customs he did not understand (Benhabib 1992: 63). Some classical myths depicted the lands and seas beyond the Mediterranean as inhabited by savages or monsters, constructions which helped to legitimise the unsympathetic way distant peoples were dealt with. Similarly, in ancient Egypt foreigners living away from the Nile were grouped with animals in some literature. In Hebrew scripture, 'You shall have but one law for the home-born and for the outsider who lives among you' (*Exodus* 12: 49), and you shall love the stranger (*Deuteronomy* 10: 18), yet in practice distinctions were common. Thus, in Judaism, 'the problem of social ethics is essentially how to transpose the mutual, caring characteristics of family-life at its best into the larger contexts of city, state, and ultimately, humanity as a whole' (Goldberg and Rayner 1989: 304).

Response to others involved emotion as well as custom, law and religious edict. In Aristotle's *Rhetoric*:

> [T]he nearness of the terrible makes men pity. Men also pity those who resemble them . . . for all such relations make a man more likely to think that their misfortunes may befall him as well . . . sufferings are pitiable when they appear close at hand.
> (quoted in Ginzburg 1994: 108)

Distance leads to indifference, while closeness can lead to pity, or to envy and destructive rivalry, part of the everyday experience of face-to-face society. The role of the emotions in ancient times is identified in 'justice personal, passionate and situated' (Solomon 1995: 256), like retribution in

Homer's epics, essentially contextual, depending on the situation and the relationships between those involved.

The situation, including spatial relations, could generate something more than a simple distinction between favoured insiders and outsiders treated differently. The wider the known world became, with the approach of modernity, the more subtle the spatial discrimination might be. Suggestions of a distance decay effect identified by such diverse figures as Confucius and Hutcheson were noted in Chapter 2. Bauman (1993: 166) makes a similar point, proposing that 'moral concern would reach its highest intensity where knowledge of the other is at its richest and most intimate, and that it would thin out as knowledge tapers off and intimacy is gradually transformed into estrangement'.

A major concern of Enlightenment thinking on morality was the supposed distinction between reason and feeling (or natural passion), inherited from classical literature. The former found its most influential expression in the abstract universalism of Immanuel Kant, who wrote in *The Doctrine of Virtue*: 'No moral principle is based . . . on any *feeling* whatsoever . . . For feeling, no matter by what it is aroused, always belongs to the order of *nature*' (quoted in Okin 1989: 231). However, feelings still engaged some of those dedicated to formalising the subject of ethics. For example, David Hume, in his *A Treatise of Human Nature*, comments on what is taken to be the self-evident character of partiality with a spatial expression:

> We find in common life that men are principally concern'd about those objects, which are not much remov'd either in space or time, enjoying the present, and leaving what is afar off to the care of chance and fortune . . . The breaking of a mirror gives us more concern when at home, than the burning of a house, when abroad, and some hundred leagues distant. (quoted in Ginzburg 1994: 116–17)

This reflects a common moral intuition recognising 'the special claim of immediacy, which makes distress at a distance so different from distress in the same room' (Nagel 1986: 176).

The kind of people involved, as well as their location, could form the basis for common-sense differentiation. As Henry Sidgwick explained in *The Methods of Ethics*, reflecting gender exclusivity as well as the racism of the times:

> We should all agree that each of us is bound to show kindness to his parents and spouse and children, and to other kinsmen in a less degree: and to those who have rendered services to him, and any others whom he may have admitted to his intimacy and called friends: and to neighbours and to fellow countrymen more than others: and perhaps we may say to those of our own race more than to black or yellow men, and generally to human beings in proportion to their affinity to ourselves. (quoted in Belsey 1992: 38)

However, Sidgwick went on to say: 'those who are in distress or urgent need have a claim on us for special kindness' (Belsey 1992: 48, note 11).

As the known world continued to expand, one of the major challenges became how to live with distant strangers some of whose moral claims could not be ignored (as explained in Chapter 2). But again, it was not simply a matter of physical distance to be overcome. Discourses of racial superiority helped to legitimise the inferior treatment of others, who could be dehumanised as clearly as the monsters of Greek myths. The expansion of European empires and the development of the capitalist world economy required the fitting of dependent territories and peoples into the cosmic order of the dominant powers: 'Beyond the spatial limits of civilization, there were untamed people and untamed nature to be incorporated . . . a spatial and cultural boundary was drawn between civilization and various uncivilized, deviant "others"' (Sibley 1995: 49–50).

Within the emerging modern city, with its anomie enhanced by spatial segmentation, the problem was one of relationships with different others living in close proximity. The American city, with its patchwork of immigrant ghettos which excited the Chicago sociologists (see Chapter 3), became the archetype of a new urban form. Familiar social space could stop at the neighbourhood's boundary, on the other side of which was alien territory, beyond familiar social norms. The internal organisation of the city was responsible for 'isolating homely spots from the wilderness in between' (Bauman 1993: 158). The strange outsiders were now within, and it took the wider threat of an exploitative economy to forge solidarity across ethnic divides.

If anything other than class consciousness, or the melting pot of cultural homogenisation, could solve modernity's problem of how to live with close as well as distant strangers, it was the impersonal rules of pecuniary transactions which came to dominate social relations hitherto conducted largely on the basis of reciprocity and trust among people familiar with one another. Part of the price was the dehumanisation of important realms of social interaction, along with the commodification of labour as an inanimate factor of production subject to the impartial evaluation of Adam Smith's 'hidden hand' of market forces.

While the philosopher's solution – universalism expressed in the Enlightenment ideal of impartiality – may have worked at a theoretical level, premodern conventions of partiality continued to prevail in many contexts: 'The fact that many universalists have in practice narrowed the scope of their principles to exclude certain others – barbarians, women, slaves, the heathens, foreigners – shows that the principles by which they actually lived have far-less-than cosmopolitan scope' (O'Neill 1996: 11, note 1). The Enlightenment discourse of reason and impartial justice, along with such other innovations as (spatially restricted) rights of citizenship, served to mask the reality of continuing spatial discrimination in the exercise of beneficence, along with some far from benevolent practices (with extensive spatial scope) associated with colonialism and imperialism as well as with capitalism. If a moral ideal of mutual concern could actually be found, it was likely to be in the localised community.

COMMUNITY AND MORALITY

The most obvious expression of the association of locality and morality is in the notion of a community. There have been interminable debates about the nature of community, within geography and other fields. However, three common elements identified in an early review of different definitions – social ties, social interaction and area (Hillary 1955) – are still reflected in a contemporary source in which community is defined as a 'social network of interacting individuals, usually concentrated in a defined territory' (Johnston, Gregory and Smith 1994: 80).

It is important to note the caveat 'usually' with respect to territory, in the context of the changing spatial scope of human interaction, and the changing nature of local social life. Knox (1995: 214) is more specific, saying that communities are defined by groups 'which may be locality-based, school-based, work-based or media-based'. Davies and Herbert (1993: 1) also recognise that the kind of human associations involved do not always occur in some defined area. This raises the question of what kind of communities exist without a local territorial base, and how they function as settings for social interaction which will, among other things, incorporate a code of morality. However, discussions linking community with morality usually proceed on the assumption of a particular locality.

An important feature missing from the descriptive definitions usually adopted in geography is that of community as a normative ideal. But even if not made explicit, much discussion of community proceeds as though this particular arrangement of human life is for the good. Sayer and Storper (1997: 8) refer to 'the honorific status of "community" in popular ideology as a warm and secure alternative to the anomie of modern society'. *Gemeinschaft*, with its connotations of moral unity, rootedness, intimacy and kinship, is preferable to the *Gesellschaft* associated with the anomie of large-scale urbanisation. Benhabib (1992: 69) sees the liberal conception of historical progress as having lost 'a coherent sense of community and a moral vocabulary which was part of a shared social universe'. To Etzioni (1995: ix), 'communities are social webs of people who know one another as persons and have a moral voice', the articulation of which can challenge the 'moral anarchy' and fill the 'moral vacuum' of contemporary society.

Two moral dimensions of community may therefore be recognised: that community is good in itself and that it speaks with moral authority. However, the quality of this good and the message of the moral voice will depend on the nature of actual human groupings which might be considered communities. This requires recognition of features of community beyond the familiar social ties, interaction and, perhaps, area, which are held to be both observable and desirable. The key variables at stake in the construction and nurture of a community have been identified as follows (Selznick 1992: 361–4): historicity (interpersonal bonds fashioned in a shared history and culture), identity (a sense of community manifest in loyalty, piety and a distinctive identity), mutuality (relations of interdependence and reciprocity), plurality (persons engaging in intermediate associations and group attachments), autonomy (the flourishing of unique and responsible persons), participation (in different roles and aspects of a society) and integration (via political, legal and cultural institutions). The extent to which reality departs from this ideal will have an important bearing on the normative evaluation of community.

Communitarianism has attracted considerable attention in recent years, as a challenge to the philosophy and practice of liberalism (see Mulhall and Swift 1996). To some, 'Liberal premises are thought to be overly individualistic and ahistorical; insufficiently sensitive to the social sources of self-hood and obligation; too much concerned with rights, too little concerned with duty and responsibility' (Selznick 1992: xi). Others add the criticism that liberalism devalues community and political participation therein, as fundamental to the good life (e.g. Buchanan 1989: 852).

As a political or moral philosophy, communitarianism has a number of strands. Three principle elements may be recognised (Kymlicka 1993: 367–70; see also Kymlicka 1990: Chapter 6). The first is that the virtues of benevolence or solidarity characteristic of communities render justice a remedial virtue (Sandel 1982), which comes into play only if community breaks down, rather than being the first virtue of social life as claimed by liberals (e.g. Rawls 1971). The second is that justice arises from particular community understandings of social goods (Walzer 1983), which are local and historically specific, rather than from external universal criteria. The third is that the common good of the community comes before individual rights, including the freedom to pursue personal conceptions of the good life.

Communitarianism recommends the maintenance or creation of community, as good in itself. The family is often regarded as the ideal case of community (evidence to the contrary, like violence within the family, notwithstanding). This involves mutual selfless generosity with no role for the notion of fair shares: in this sense, beyond justice. Sandel (1982: 31) notes a range of less extreme cases, including universities, trade unions, tribes, nationalisms and various ethnic, religious, cultural and linguistic communities. While he also includes neighbourhoods, cities and towns, his communities are not all territorially defined.

Central to communitarianism is a particular notion of human being. This is sometimes described as a relational identity or social self, expressed as a focus on a localised 'we' rather than an autonomous 'I': 'one of us' could be interpreted as 'part of us'. Thus:

> [T]o say that members of a society are bound by a sense of community is [to say] that they conceive their identity – the subject and not just the object of their feelings and aspirations – as defined to some extent by the community of which they are a part. For them, community describes not just what they *have* as fellow citizens but also what they *are*, not a relationship they choose (as in a voluntary association) but an attachment they discover, not merely an attribute but a constituent of their identity. (Sandel 1982: 150)

Such a conception goes beyond the recognition of community as instrumental in the pursuit of individual ends, and as subject of purely sentimental attachment. Persons are seen as embedded or situated in a particular geographical and historical milieu, in contrast to the autonomous self-determination of agents floating free of societal context in the idealised liberal formulation.

This conception of human identity is closely associated with the small-scale premodern society (to which reference was made in Chapter 2):

> In many pre-modern, traditional societies it is through his or her membership in a variety of social groups that the individual identifies himself or herself and is identified by others. I am brother, cousin and grandson, member of this household, that village, this tribe. These are not characteristics that belong to human beings accidentally, to be stripped away in order to discover 'the real me'. They are part of my substance, defining partially at least and sometimes wholly my obligations and my duties. (MacIntyre 1985: 33-4)

Persons are thus identified through roles which bind them into communities in and through which specific goods are attained; the egoist in the ancient and medieval world was someone who had 'excluded himself from human relationships' (MacIntyre 1985: 229). Something of the attraction of the traditional community is captured as follows: 'What it offers is human warmth and entanglements along settled paths rather than the poignancy of the chance encounter, at an unfamiliar place, with a stranger' (Tuan 1986: 19).

Communitarianism has recently (re)surfaced as a political project with explicit moral content. On both sides of the Atlantic and at different points on the political spectrum, there are calls for a return to values associated with community, loss of which is thought to be implicated in various ills. Whether a result of state-sponsored dependency or market-induced competitiveness, the moral fabric of communities is thought to be disintegrating, causing family breakdown, criminality and social disorder (Bowring 1997: 96). Hence the proclamation of 'a social movement aiming at shoring up the moral, social, and political environment' (Etzioni 1995: 247). More specifically:

> Communitarians call to restore civic virtues, for people to live up to their responsibilities and not merely focus on their entitlements . . . Communities draw on interpersonal bonds to encourage members to abide by shared values . . . Communities gently chastise those who violate shared moral norms and express approbation for those who abide by them. They turn to the state (courts, police) only when all else fails. (Etzioni 1995: x, ix)

The normative content of the notion of community gives communitarianism a special appeal, compared with the more threatening conservatism, socialism or communism. Part of the political attraction is that supposed (even imagined) communities can be conveyed as an existing resource, to be deployed in such strategies as 'community care', thus relieving the state of what might otherwise be substantial expenditure. Hence the apparent paradox of the invocation of communitarian values by governments otherwise committed to liberal individualism.

The debate between communitarianism and liberalism may be traced back to the challenge to Kantian formalism posed by Hegel's notion of the ethical community (*Sittlichkeit*) rooted in ongoing social relationships, shared practices and traditions. Hegel is sometimes credited with seeking 'a middle ground between the excesses of a political theory based solely on the idea of rational autonomy and one based exclusively on the ideal of historically sedimented community' (Stern 1991: 261). Marxism assigns value to community and to individual autonomy, but sees them both threatened by the exploitative social relations of capitalism (see e.g. Peffer 1990). Communitarianism has become a haven for some who reject both capitalism's rampant individualism and socialism's imposed conception of the good, and seek a 'third way'.

Communitarianism and liberalism have positive as well as normative aspects, making claims about what is true as well as about what is good. Walzer (1990: 7–10) notes two different and contradictory arguments on the positive side. The first is that liberal theory accurately represents its practice: contemporary Western societies are 'the home of radically isolated individuals, rational egotists, and existential agents, men and women protected and divided by their inalienable rights'. In other words, communitarians are wrong about the way the world actually is. The second argument is that liberal theory misrepresents real life: 'It is in the very nature of a human society that individuals bred within it will find themselves caught up in patterns of relationships, networks of power, and communities of meaning.' In short, communitarians are factually correct.

However, observation suggests that strict communitarianism and liberalism represent extremes or ideal types, both of which contain elements of truth. As Walzer (1990: 11–18) recognises, we do live in an unsettled and mobile society, yet we are still 'creatures of community' with some common values, and with communal feelings and beliefs more stable than we once thought they were. And on the concept of the self, which is supposed to divide communitarianism and liberalism: 'Contemporary liberals are not committed to a presocial self, but only to a self capable of reflecting critically on the values that have governed its socialization; and communitarian critics, who are doing exactly that, can hardly go on to claim that socialization is everything' (Walzer 1990: 21). As to the homogeneity and common values supposed to characterise community, 'a proper understanding of community, from a sociological point of view, presumes diversity and pluralism as well as social integration' (Selznick 1992: xi).

In addition to its challenge to liberalism, communitarianism may be viewed in terms of opposition to modernity. The move (literally) from cohesive small-group social ties mediated by relations of reciprocity to the impersonal ties based on efficiency mediated by contractual relations was closely associated with the modern phenomena of capitalist industrialisation and urbanisation. So, when some communitarians advocate a return to community as the embodiment of human good, it is to the premodern world that they allude. For instance looking for an alternative to modern liberal individualism and socialist collectivism as a model for postapartheid society may invoke the relational identity of traditional African communal sentiments:

> [T]he more fully I am involved in community with others the more completely I am able to realise my own deep desires to the full. The good of the community (in which I am also involved) will be my highest value, just as traditional African thought would expect. At the same time, the influence of the community on me is what enables me to achieve this form of self-transcendence and self-donation which is the fullest expression of my own self-realisation. (Shutte 1993: 90)

It requires only cursory familiarity with the values of contemporary South African society to appreciate the challenge involved in recreating these premodern ethics (see Chapter 8).

However, the philosophical (rather than practical) point remains. There is something attractive about these earlier ethics, as the modern era moves into what some describe as postmodernity, with its fragmentation and uncertainties, and its threat of moral relativism or nihilism. This has much to do with the mutuality and care for one another often associated with the traditional community: 'When the term *community* is used, the first notion that typically comes to mind is a place in which people know and care for one another' (Etzioni 1995: 31).

AN ETHIC OF CARE

Until quite recently, moral philosophy was largely a male (even masculinist) endeavour. What was considered good or right reflected a male view of the world, as a life of rational public discourse conducted by persons unencumbered by the social relationships which might impede dispassionate thought, consistent with some versions of liberalism. The place of women was the private realm of the home, beyond the reach (and protection) of justice, as in the communitarian ideal.

The possibility of gender differences in moral thinking was first provoked by Lawrence Kohlberg (1981), who proposed a theory of moral development in sequential stages. These began with simply avoiding punishment, and culminated in a commitment to fairness guided by universalisable and impartial moral thinking. Kohlberg's findings were challenged by Carol Gilligan (1982), whose work with female respondents suggested that they were more concerned with relationships than with rights and rules. She therefore proposed that women articulate morality in a different voice, contrasting an ethic of care with that of justice:

> In this conception, the moral problem arises from conflicting responsibilities rather than from competing rights and requires for its resolution a mode of thinking that is contextual and narrative rather than formal and abstract. This conception of morality as concerned with the activity of care centers moral development around the understanding of responsibility and relationships, just as the conception of morality as fairness ties moral development to the understanding of rights and rules.
> (Gilligan 1982: 19)

The full Kohlberg–Gilligan controversy is summarised elsewhere (Kymlicka 1990: 262–86; Benhabib 1992: Chapter 5; Tronto 1993: Chapter 3; Hekman 1995: Chapter 1). It is sufficient here to note that the ethic of care came to be seen as a supplement (some would say alternative) to the dominant moral tradition of the Enlightenment, rather than a distinctive and exclusively female voice challenging the male voice of impartial justice as the peak of moral refinement. 'Care ethics raises caring, nurturing, and the maintenance of interpersonal relationships to the status of foundational moral importance' (Friedman 1993: 147), such that gestures in this direction can now be found in works otherwise devoted to impartiality (e.g. Barry 1995: 246–55). Care might be regarded as the most general, and universal, individual need – from cradle to grave:

'humans are not fully autonomous, but must always be understood in a condition of interdependence . . . all humans need care' (Tronto 1993: 162). There may be specific forms of human relationships and societies that are particularly effective in caring; thus, 'some of the basic needs of each individual are to belong to a community, to be recognized, to share, to care, and to be cared for' (Markowic 1990: 132).

There is an emphasis in some feminist writing on the importance of knowing the other for whom we care, in the active sense of *for* rather than *about* (Clement 1996: 16–17). Hence the view that caring for distant people is care in name only: we cannot care for people we do not know (Noddings 1984). Some persons may be better placed than others (literally, in a geographical sense) to care for those in need: 'being uniquely situated to answer to someone's needs derives from an ongoing relationship . . . I am in frequent proximity of her, and I know her needs and desires better than you do' (Friedman 1991b: 822). There is reference to 'a kind of caring that requires knowing people in their concrete particularity rather than as representatives of certain disadvantaged groups' (Jagger 1995: 132).

The case for this kind of partiality in care is by no means confined to feminism. There is consequentialist, utilitarian argument for favouring nearest and dearest, given the (impartialist) requirement that everyone carries the same weight in the aggregation of wellbeing:

> It is easier to know what people nearby need, and how best we can help; it is easier to get the necessary aid to them efficiently, without losing too much in the process . . . it makes a good deal of utilitarian sense to assign particular responsibilities for particular people and projects to other people near to hand. (Goodin 1991: 246–7)

Because we know more about close people, the probability of doing good in an uncertain world is greater:

> Achieving good results is very often a matter of coordinating a series of actions rather than scattering largesse around . . . the good consequentialist should focus her attentions on securing the well-being of a relatively small number of people, herself included, not because she rates their welfare more highly than the welfare of others but because she is in a better position to secure their welfare.
>
> (F. Jackson 1991: 475, 481)

However, this argument is disputed by those who hold that one would do more good by devoting one's resources to famine relief rather than to friendship, in circumstances of neutrality between recipients (Cocking and Oakly 1995: 94–5, note 14). If the argument from consequentialism does carry, then relaxing neutrality in favour of nearest and dearest adds further strength to their claims.

There are obvious links between the ethic of care and communitarianism. Both involve mutuality and a relational rather than autonomous identity. Both reflect something of premodern ethics, as in the connections sometimes made between the ethic of care and traditional African morality (Tronto 1993: 83–4; see Chapter 8). And both imply partiality, favouring members of our own family, group or community. What is more, they both appear to rely heavily on the proximity of the persons concerned, to make their morality work.

LIMITATIONS OF LOCALITY, COMMUNITY AND PARTIALITY

So far, we have not problematised the ideals attributed to community and to the ethic of care, including the partiality associated with locality. It is to elements of critique that we now turn.

First to communitarianism, which is by no means a homogenous philosophy. Veit Bader (1995) distinguishes between 'conservative' or 'protective' communitarianism and a 'liberal-democratic' version or 'communitarian liberalism'. The first of these relates to more traditional premodern communities, of the kind invoked by MacIntyre (1985). The second refers to what Hekman (1995: 58) describes as a dialectical concept fusing elements of liberal individualism with communitarianism, of the kind that might actually exist in many modern societies, sometimes portrayed as 'community liberated' (Davies and Herbert 1993: 28).

The conservative version is the subject of various myths:

> [I]nternal homogeneity of communities is postulated and cross-cutting communal allegiances and collective identities are forgotten; in a kind of retrospective nostalgia, communities are thought to be harmonious (traditional) *Gemeinschaften* and confronted with conflict-ridden (modern) strategic *Gesellschaften*; cultural communities are constructed without any analysis of structural antagonism and conflict, particularly class antagonism and conflict; the idea of shared meaning, of shared cognitive and normative frames and interpretations is very much overstressed. (Bader 1995: 217)

Traditional communities were (and, insofar as they survive, still are) often characterised by various forms of oppression, protecting the prevailing value system including its moral code. This constrains individual autonomy and freedom to choose a way of life, leaves little room for criticism and change, and may sustain structural inequality involving status hierarchies facilitating class exploitation and patriarchal domination of women. Civil rights may be restricted, as may freedom to enter and leave the community. Hence the response of some liberals:

> Without the protection for autonomy and independence guaranteed by liberal rights, the individual, absorbed in community, unable to reflect critically upon her role, her obligations, and the character of her community as a whole, may become an unwitting accomplice in an immoral way of life. (Buchanan 1989: 871–2)

Overlooking these kind of realities results in idealisation. Hekman (1995: 55) notes that 'a disturbing characteristic of much communitarian literature [is] the romanticization of premodern societies that ignores the oppression and hierarchy endemic to them'. For example, the traditional African societies invoked above were brutally repressive of internal dissent or challenge to the authority of tribal chiefs. And the treatment of women leads some to find communitarianism a perilous ally for feminist theory (Friedman 1991a: 305; see also Hekman 1995: 58–62).

Idealisation may also be involved in accounts of what are often considered models of modern communities. For example Jocelyn Cornwell (1984: 52–3) claims that the urban village community in East London celebrated by M. Young and P. Wilmott (1957) was presented to them in this way by local residents – against a background of rapid social change – as an image of a lost way of life. Contrasting with this were accounts by Cornwell's own respondents revealing a past of economic distress and of brutality, especially by men on women, which hardly evoked the qualities of warmth and sympathy associated with the ideal community. Thus: 'Slides between descriptive and normative uses of "community" tend to conceal the divisions of the former behind the harmony of the latter' (Sayer and Storper 1997: 8).

A common criticism of the traditional conception of community is its intolerance of difference. 'The ideal of community . . . denies the difference between subjects. The desire for community relies on the same desire for social wholeness and identification that underlies racism and ethnic chauvinism on the one hand and political sectarianism on the other' (I. M. Young 1990a: 303; see also 1990b). East London again provides an example:

> There is a strong sense of community in Bethnal Green, but it should be noted that where there is belonging, there is also not belonging, and where there is in-clusion, there is also ex-clusion. In East London, the dark side of community is apparent in a dislike of what is different, which finds its clearest (but by no means its sole) outlet in racial prejudice. (J. Cornwell 1984: 53)

Homogeneity, if not perfect harmony, was part of what had been lost, as people looked back. If the good of mutuality was manifest in the traditional community, its moral voice may have articulated partiality and prejudice which continues to impede the inclusion of some 'others' (especially immigrants from Bangladesh) in a strongly differentiated part of an increasingly multicultural society.

'Liberal-democratic' communitarianism escapes some of the criticism of the conservative version. There is room for limited individual freedom and internal critique. But this kind of communitarianism tends to overlook problems of external relations among local communities. These can take the form of parochial closure and power-asymmetries between communities: domination, oppression, and exclusion (Bader 1995: 222).

Closure involves selectivity of community membership. Thus:

> The primary good that we distribute to one another is membership in some human community . . . it is only as members somewhere that men and women can hope to share in all the other social goods – security, wealth, honor, office, and power – that communal life makes possible. (Walzer 1983: 31, 63)

Communitarians therefore tend to defend boundaries, to protect the homogeneity of culture, values and so on, whereas under liberalism 'boundaries of the group are not policed; people come and go, or they just fade into the distance' (Walzer 1990: 15). By constraining population movement, closure can also become a means of protecting local privilege and perpetuating the inequality among communities which generates power differentials. The moral case for closure may therefore depend on reducing the degree of inequality among and within communities: 'closure under conditions of "rough" equality differs radically from closure under conditions of systematic exploitation, oppression, discrimination, and exclusion' (Bader 1995: 219).

The charge of parochialism, to which communitarianism is prone, can also be applied to the ethic of care. Favouring our nearest and dearest seems a natural human sentiment, perhaps intrinsic to the process of care itself. The strength of obligation as well as of desire to care for others may depend on where as well as who they are. We may also be able to care for close people more effectively than those at a greater distance, as was suggested above. However, there is a risk of over-prioritising the mutuality of face-to-face relationships, which is also found in communitarianism. Critics of parochial readings of an ethic of care have provoked the concession that we might construct ever widening circles of care, and urge those nearby the distant needy to care for them (Clement 1996: 17; see also Friedman 1993: Chapter 3).

Returning to the practical issue, it is debatable to what extent close familiarity with the needs of others is actually required for care, whether to recognise the specific needs or to meet them most effectively. P. Singer (1972: 26) concedes that we may be in a better position to judge what needs to be done to help a near person than one far away, and perhaps also to provide the assistance necessary, but instant communication and swift transport now enables aid to be disseminated, with the assistance of expert observers and supervisors: 'There would seem, therefore, to be no possible justification for discriminating on geographical grounds.' While everyday experience confirms that the needs of family members may be first identified within the family, and that some kinds of response may depend on such proximity, the diagnosis of need and the provision of effective care can seldom be so confined. However, the real danger of outside aid agencies or professionals imposing inappropriate remedies suggests that some insider knowledge of the situation is helpful. This leads to broader issues concerning the distribution and institutionalisation of the means of caring, including the importance of linking care with a theory of justice, to which we will return in the next chapter.

A PREMODERN COMMUNITY: THE SHTETL

This chapter concludes with a brief portrayal of a premodern community. The example chosen is the 'shtetl' (small town, in Yiddish), places of residence for most of the Jewish population of Poland, and some other parts of east-central Europe, who were not living in the larger cities. The shtetl illustrates aspects of the ambiguity of the traditional community, its representation and interpretation. Part of the shtetl's particularity or difference was a distinctive setting, for it was embedded in a specific social and political context crucial to its life, and to its eventual death. Visiting the shtetl enables us to return to aspects of the Jewish experience, introduced in Łódź at the end of the previous chapter, and continued as codas to the two chapters which follow.

Poland began to attract Jews in significant numbers from the thirteenth century onwards. Their predominantly urban settlement led to a stable population, engaged in a range of economic activities. A form of local autonomy enabled the Jews to retain and develop their own culture, separate from the Polish people, but by no means in isolation for there was much commercial interaction. Small towns with their Jewish populations proliferated in the rural landscape; by the mid-eighteenth century a quarter of the world's Jewry lived in the Polish shtetl (Steinlauf 1997: 5). But by the beginning of the present century economic decline had set in, and by the 1920s much of the population had migrated to the cities, such as Łódź, in search of work.

The topography of the shtetl was characterised by permanent and repetitive features: the market square, synagogue, inn, shops, winding streets and Jewish cemetery (for further details see Bar-Gal 1985). But 'shtetl' came to refer to a cultural as well as a physical entity, representing the *modus vivendi* of Jews in Eastern Europe, a bastion of traditional Jewish culture, for which the term itself became a synonym (Prokopówna 1989: 129). In Poland Jews were able to create a particularly coherent culture, characterised by what has been described as 'the highest degree of inwardness' (A. Heschel, quoted in Steinlauf 1997: 4). Such was the ubiquitous presence of the Jews that a nineteenth-century novelist could ask, with ominous prescience: 'And do you know what makes every town Polish? The Jews. When there are no more Jews, we enter an alien country and feel, accustomed as we are to their good sense and service, as if something were not quite right' (J. I. Kraszewski, quoted in Prokopówna 1989: 132–3).

Something of the shtetl as a community is captured by Eva Hoffman (1998), in a book on the history of one such place – Brańsk (mentioned in the previous chapter). The shtetl was a resilient social formation, deeply religious and traditional:

> Piety provided the order both of concrete daily rituals in which everyone participated, and of seemingly eternal verities and values by which everyone was guided. Perhaps the main virtue of the shtetl for its inhabitants was the extent to which it was a community – small, closely interwoven, reassuringly familiar.
> (Hoffman 1998: 12)

There were two sides to the relative closure of the shtetl, and the similarity among these settlements:

> Each shtetl was a self-contained world, and each was utterly recognisable as an instance of its kind. This consistency, the patterned predictability of life, was undoubtedly part of the shtetl's strength. But it also meant that the shtetl was a deeply conservative organism, resistant to innovation, individuality, or rebellion.
> (Hoffman 1998: 91)

So for the restless, inquisitive and nonconformist, the shtetl appeared insular, superstitious and opposed to progressive trends.

The shtetl was both inclusive and exclusive, in senses common to many other premodern communities:

> The dense web of communal associations embraced just about everyone. No one was left out – and no one, for that matter, was allowed to escape it. Even the poorest and most improvident members of the community were included in the communal net . . . These institutions gave the shtetl a kind of horizontal coherence, but the social map was also diagrammed vertically, by professional status and other forms of hierarchy. (Hoffman 1998: 94)

Divisions between the rich and the poor were evident, yet to some extent subsumed under the dominant system of virtues within which the religiously learned man was valued above the merely wealthy. Religion permeated the fabric of a life which was otherwise pragmatically materialistic:

> The day, the week, and the year were shaped and parsed by ritual signposts . . . Each part of life, from food to sex to marriage and personal hygiene, was governed by a highly elaborate and precise body of religious principles and rules.
> (Hoffman 1998: 97)

The strict gender roles typical of premodern societies were given special character and meaning by interpretation of religious texts. It was from these that rabbinical authority was able to tease out a particular version of moral truth, albeit subject to debate among men with access to the arena within which scholarship flourished as a virtue. Otherwise, wisdom was, literally, received.

An important distinguishing feature of the Jewish community was its practical commitment to an ethic of care. Polish Jewry organised itself into what has been described as the most effective autonomous system of government in the Diaspora (Goldberg and Rayner 1987: 118). Each community chose a body of learned (male) elders, chief of whom would be the rabbi, responsible for running local affairs, including maintenance of the population's educational and social requirements, adjudication of legal and doctrinal disputes, and collection of taxes. For much of the centuries of the shtetl's existence, a national body composed of leading religious and lay members coordinated matters of Jewish concern, ensuring that communal affairs ran as smoothly as turbulent times permitted.

Michael Walzer (1983: 71–4) features the medieval Jewish community as a case of goods customarily provided on the basis of membership. The major form of general provision was religious, with the synagogues and courts paid for out of public funds, but also included services which would now be regarded as secular. The community provided public baths, and supervised the slaughterers responsible for kosher meat. Some effort was made to keep streets clear of rubbish, and to avoid overcrowding neighbourhoods. Towards the end of the medieval period, many communities provided hospitals, doctors and midwives, well before their availability in many other places. Much attention was paid to educational provision: closely allied to religion, this was crucial to the reproduction of community values. Financial or material support for the poor was central to this ethic of care. The poor of one's own community took precedence over other Jews: to this extent beneficence was localised. Contribution was a social and religious obligation arising from functional necessity, rather than an act of altruism: 'It was hardly possible to live in the Jewish community without contributing; and short of convertion to Christianity, a Jew had no alternative; there was no place else to go' (Walzer 1983: 72).

After centuries of relative stability, interrupted by the occasional pogrom or war, the shtetl was yielding to modernisation as it entered the twentieth century. The kind of social formation which might have emerged by a process of evolution will never be known. What survived the Holocaust is selective recollection, supplemented by such records and reconstructions as zealous scholarship and creative imagination has made possible (see for example the depiction of a day in the life of a Polish-Jewish town, in Szewc 1993). The result has been a selective representation, or romanticisation, of shtetl life itself.

The shtetl provides a poignant case of the idealisation of the traditional community, referred to in the previous section of this chapter. For example the revival of the Kazimierz district of Kraków (Figure 4.1), with its Jewish restaurants and Judaica for tourists, has been portrayed by Anne Karpf (1996: 303–4) as part of a sentimental memorialisation of the shtetl, reaching its apogee in the musical *Fiddler on the Roof.* She depicts prewar Jewry as not a homogeneous world of the devout, but richly diverse, fractious, obscurantist as well as scholarly, often dreadfully anti-feminist, including socialists as well as the assimilated and partly assimilated, and also with great poverty and hardship. She criticises the celebration of 'a timeless and idealised Eastern European shtetl which didn't square with the pre-war reality – the shtetl wasn't a cultural island but far more heterogeneous and changing' (Karpf 1996: 341, note 8). Similarly, Rabbi Julia Neuberger (1996: 259) associates a renaissance or rediscovery of Jewish culture with a faulty re-invention of the romantic idea of the shtetl without recognising that it was 'dirty, dangerous, and very, very poor'.

The usual description of the shtetl in Polish literature was as ruined, neglected, and wretchedly poor. The hero of a novel published in 1925 saw 'the quagmire of muddy streets . . . houses, all of various shapes, colours and degrees of filthiness . . . pigsties and puddles . . . farm buildings and burnt rubble . . . the market place surrounded by Jewish shops, their doors and windows splashed with months-old mud, and unwashed' (Prokopówna 1989: 132). There could be an element of anti-Semitism in the repeated images of dirt, subsequently to be cleansed or purified (in the manner suggested by Sibley 1988). But a different image was being conveyed by Polish-Jewish writers in the inter-war years of the erosion of shtetl life: as 'a familiar and sympathetic entity, a happy and secure place, despite its aesthetic shortcomings . . . the affectionate abode of memory and imagination, an intimate entity . . . a universal landscape, as it were, of Jewish biography' (Prokopówna 1989: 135). It is not therefore surprising that, when decay was replaced by annihilation, this tone of nostalgic idealisation was sometimes exaggerated in writings of Holocaust survivors. The rhetorical heights took on a metaphysical tone: 'hundreds of little towns were like holy books . . . The little Jewish communities in Eastern Europe were like sacred texts open before the eyes of God' (A. Heschel, quoted in Prokopówna 1989: 130).

There is a link to be made with the idealisation of another archetype community, that of East London, referred to in the previous section. In her interpretation, J. Cornwell (1984) drew on an analogy between loss of community as a result of slum clearance and the experience of bereavement and mourning (Marris 1974). Whereas societies have their own ritualised ways of coming to terms with human death, this is not the

Figure 4.1 *The Kazimierz district of Kraków preserves typical feature of the traditional Jewish community of the ghetto or shtetl, but in a changing context.* Top: *the Old Synagogue (left) and entrance to the cemetery (right), with (post)modern insertion reflecting the district's rediscovery and revival.* Bottom: *Jewish restaurant, part of the romanticisation and commercialisation of Polish Jewish heritage (see above, p. 72). (Source: Author's collection)*

case with loss of home and neighbourhood, which can make this experience so hard to bear. People can get stuck in the mourning process, over-valuing what may not have been a particularly pleasant environment. This helps to explain what she found in East London, as people looked back and idealised a former community life which was actually one of poverty and violence: 'This is chronic grief, in which the bereaved person refuses to let go of the past, but holds onto it and mummifies it and romanticizes it instead' (J. Cornwell 1984: 52). The experience and recollection of the Holocaust, involving large-scale loss of persons as well as place, may have worked in a similar way.

A final point raised by the shtetl concerns its external relations: often overlooked in consideration of the community. The shtetl was as a place where Jews lived side by side with the local population:

> Polish shtetls were usually made up of two poor, traditionalist, and fairly incongruous subcultures: Orthodox Jews and premodern peasants. Morally and spiritually, the two societies remained resolutely separate, by choice on both sides. Yet they lived in close physical proximity and, willy-nilly, familiarity. In the shtetl, pluralism was experienced not as an ideology but as ordinary life.
>
> (Hoffman 1998: 12)

As Jews in Poland lived remote from old centres of rabbinical authority, relaxation of religious rigour may have made it possible for Christians and Jews to live together without the rigid barriers that obtained elsewhere, perhaps easier to retain fluidity of individual self-definition and to cross borders between group identities (Hoffman 1998: 34–5). There was certainly regular interaction among the two populations:

> When the market was a weekly meeting place for Poles and Jews; when a Polish farmer haggled with a Jewish merchant and they arrived at a price; when the same Polish farmer later went to relax at an inn run by a Jewish lease-holder; then archetype might have been modified by actual contact . . . some consciousness of real persons, of men and women who were more or less amiable, witty, attractive, wise, honest, or strong. (Hoffman 1998: 74)

This might have been expected to generate some of the mutuality, reciprocity and trust on which intra-community relations of solidarity are based. But there was another dimension to the context, reflecting a long history of different and separate identities, punctuated by periods of overt animosity, hatred and violence. Thus:

> The villages and small towns were where Jews and Poles were at their most exposed and vulnerable, and where ongoing political conflicts were at their sharpest. This was where Jewish inhabitants experienced acts of the most unmediated cruelty from their neighbors – and also of most immediate generosity. In the dark years of the Holocaust, the shtetl became a study in ordinary morality tested, and sometimes warped, by inhuman circumstances. (Hoffman 1998: 13)

Something of this will be explored in the next chapter, as a conclusion to consideration of the moral significance of proximity, and of distance.

CHAPTER 5

DISTANCE: THE SCOPE OF BENEFICENCE

> Whether we can conceive of a way to think of morality that extends some form of sympathy further than our own group remains perhaps the fundamental moral question for contemporary life. (Tronto 1993: 59)

There is a prevailing sense of regret at loss of community, its moral limitations notwithstanding. The re-creation of community, or at least its laudable qualities, is a widely held aspiration. Life would surely be better if the concern for persons in close proximity conventionally associated with community could be transformed into a spatially extensive beneficence, in the sense of actively caring for more distant others in need.

We live in a world of expanding spatial relations. But the globalisation of modernity and market forces carries no guarantee that the benefits will be widely distributed. There is certainly a capacity to harm others in distant places, arising from the downside of economic interdependence and some highly mobile sources of pollution. But, could we conceive of a globalisation of advantage, manifest in responsibility of those better-off to care for less fortunate others elsewhere? Could the 'global village' become a moral community? Or is the parochialism of localised self-interest joining the sanctity of market outcomes as a supreme universal value? The nature of social relationships in geographical space has some bearing on these questions, for it could be crucial to the inclination as well as the capacity to care. This chapter shifts perspective from proximity to distance, to explore the spatial scope of beneficence (continuing the integration of arguments in D. M. Smith 1998b and 1999a, begun in Chapter 4).

RECONSTITUTING COMMUNITY

The first possibility to be considered is that community could be reconstituted, to take advantage of its moral strength while minimising its weaknesses. Some attempts will be reviewed briefly.

One of the most influential reassertions of communitarianism in recent years is Amitai Etzioni's appeal to 'the spirit of community' (1995).

Conventional partiality is reflected in his view that the first moral responsibility is for people to help themselves, followed by care for those closest including kin, friends, neighbours and other community members. However, 'one of the gravest dangers in rebuilding communities is that they will tend to become insular and indifferent to the fate of outsiders' (Etzioni 1995: 146). He recognises that overcoming parochialism requires an expansion of moral claims and duties, involving a responsibility to help communities whose ability to help their own members is limited. This should take place within an overarching 'community of communities', characterised by 'pluralism-within-unity', but sharing the core values of democracy, individual rights and mutual respect.

However, he claims that these values are Western, and that those committed to them 'will find little comfort in other major cultural traditions' (Etzioni 1995: 159). It therefore seems difficult to accommodate difference associated with the multiculturalism which increasingly characterises 'Western' countries. For example how would this brand of communitarianism respond to immigrants or other 'minority' populations wishing to preserve a distinctive (non-Western) culture? Is help for such communities dependent on their endorsement of a set of national core values? There is a danger here that local partiality, even chauvinism, is merely displaced to the scale of the nation state.

The 'communitarian democracy' espoused by Philip Selznick (1992) promises a more inclusive, and extensive, reconstitution of the traditional notion of community. He distinguishes between particularism as bounded altruism and universalism as inclusive altruism. 'As opportunities for cooperation are enlarged and their benefits perceived, the application of altruism is no longer limited to a small band of close relatives. Particularism is diluted as the community expands' (Selznick 1992: 195). Hence the quest for community that looks outwards rather than inwards:

> The conventional forms of particularism – bonds of family, friendships, ethnicity, and locality – need not be perceived as its only province. The same ethos may be applied, with due respect for context, to wider worlds of work, education, and government. (Selznick 1992: 206)

This expansion of community is facilitated by the recognition of human sameness – that human beings are alike in morally significant ways (see Chapter 7), without denying particular groups the autonomy to express their differences. The ideal of community as both particularist and universalist is summarised as follows:

> The experience of community begins with the very concrete affinities of early life, especially families, peer groups, and, in many cases, supportive subcultures. Taken

> by themselves these affinities are isolated and parochial. To sustain community as a framework for the whole of life and for the flourishing of multiple groups, a transition must be made from piety to civility, and from bounded to inclusive altruism. (Selznick 1992: 521)

His central thesis is that a proper understanding of community, from a sociological point of view, reveals diversity and pluralism as well as integration. The practical, political task is to realise these qualities.

Various proposals have been made for the active creation of different forms of community. Marilyn Friedman (1991a: 312) points out that, for the child (like the premodern person), the community of origin is 'found, not entered; discovered, not created', but that this need not be true of adults for whom there are now communities of choice, rather like friendships. Communities of place are of diminishing importance in urban areas, where residents tend to form social networks from people brought together for reasons other than proximity of residence. Indeed, the diversity of urban life may encourage people to link in a number of different communities of choice rather than of circumstance, based on common activities, interests or causes. This strategy appeals to feminist critics of communities containing social roles and structures oppressive towards women:

> [U]rban communities of choice can provide the resources for women to surmount the moral particularities of family and place which define and limit their moral starting points . . . these more voluntary communities may be as deeply constitutive of the identities and particulars of the individuals who participate in them as are the communities of place so warmly invoked by communitarians.
>
> (Friedman 1991a: 316)

They may encourage critical reflection on, and resistance to, the values of the original community of place. However, such a vision lacks a sense of how these various communities of choice reach out, relating to one another.

Iris Marion Young (1990b) has proposed a normative ideal of city life, as an alternative both to the conventional notion of community and to liberal individualism. She recognises that social life is structured by vast temporal and spatial networks among persons, so 'nearly everyone depends on the activities of seen and unseen strangers who mediate between oneself and one's associates, between oneself and one's objects of desire' (I. M. Young 1990b: 237). The exercise of individual freedom leads to group differentiation with a social and spatial expression, to the formation of affinity groups but without closure in the sense of inclusion and exclusion. Groups overlap and intermingle: 'In the good city one crosses from one distinct neighborhood to another without knowing precisely where one ended and the other began' (I. M. Young 1990b:

238–9). There should be equality of groups, along with mutual recognition and affirmation of their differences (I. M. Young 1990b: 191).

These and other contemporary scenarios involve a significant shift from the traditional localised or place-based community towards the community without propinquity. The 'place' is seen as increasingly open, or porous, its populations exhibiting hybrid identities. Thus, some people now experience diverse others and their cultures locally, within the increasingly cosmopolitan metropolis, selectively experiencing new art-forms, cuisines, commodities and friendships. The intimate, exclusive togetherness of local community affiliation may be giving way to a less defensive and more outward-looking sense of place: 'Relationships can be maintained over greater distances, with greater degrees of intimacy, rather than being episodic or transitory as in the past' (Davies and Herbert 1993: 16). However, the occasional longing for the cohesion of traditional communities is a sign of 'the geographical fragmentation, the spatial disruption, of our times' (Massey 1991: 24). It is part of what some people miss, and the loss of which they mourn, in an increasingly ambivalent (or confused) relationship with place as an arena of encounter as well as a container.

The importance of proximity to community thus remains an open question. For example Etzioni's new communitarianism reveals uncertainty about locality. He states that it is impossible to create a satisfactory community 'apart from geography', but notes that people who work in the same place spend longer together and in closer proximity than those who live on the same street, and that 'nongeographic communities' often provide some elements of the communitarian nexus and hence the moral infrastructure essential for a civil and humane society (Etzioni 1995: 122). Against the community without propinquity are those who argue that a revived form of place-based community may be one of the ways in which people can recognise that self-fulfilment requires moral demands beyond the self, helping to counteract some of the deficiencies of contemporary society (Davies and Herbert 1993: 27, 188). To MacIntyre (1985: 263): 'What matters at this stage is the construction of local forms of community within which civility and the intellectual and moral life can be sustained through the new dark ages which are already upon us'.

While some of these aspirations may reflect a disciplinary dedication to place or nostalgic yearning for the past, there is also an understanding that some things can still best be achieved locally. However, the practicality of (re)creating local community in a globalising world seems increasingly remote. And, while there are indications of the possibility of spatial extension of the beneficence traditionally associated with community, this might more effectively be accomplished through an ethic explicitly dedicated to care.

EXTENDING THE SCOPE OF CARE

Everyday experience suggests that favouring our nearest and dearest is a natural human sentiment, for which there are good prudential as well as moral reasons. This raises the question of how far humankind is capable of extending these relations of care to 'outsiders'. It has been customary in Western moral thought, particularly since Kant, to see a rupture between the public virtue of justice and the private virtue of goodness, but to argue that this can be mediated by extending the sympathy we naturally feel towards those closest to us onto larger human groups (Benhabib 1992: 140). The problem is that of translating this ideal into practice, with very wide if not universal scope, in a world in which local relationships remain strong binding forces in human identity and solidarity.

Joan Tronto (1987) recognises that an ethic of care could become a defence of caring for one's family, friends, group or nation, and indeed of any set of conventional relationships. Advocates of an ethic of care therefore need to consider how to spin their web of relationships widely enough that some people are not beyond its reach. Hence: 'Whatever the weakness of Kantian universalism, its premise of equal moral worth and dignity of all humans is attractive because it avoids this problem' (Tronto 1987: 661). She claims that the central questions of current moral theory are about 'how to treat morally distant others who we think are similar to ourselves'; thus, only when we extend our 'moral boundaries' to include care will we be able to deal with the implications of tribalism and racism which undermine common responsibility (Tronto 1993: 13, 59). Crucial to such a project are the spatial relationships within which people actually live, and experience human similarity as well as difference.

Various processes subsumed under the concept of globalisation have greatly changed the spatial organisation of human life in recent decades, and with it the ways in which people in different parts of the world have come to interact with one another. Understanding how 'we', in the affluent parts of the world, impact on the lives and environments of distant others, can lead to an extension of a sense of responsibility:

> To the extent that we can show that our lives are radically entwined with the lives of distant strangers – through studies of colonialism, of flows of capital and commodities, of modern telecommunications and so on – we can argue more powerfully for change within the global system ... there is no logical reason to suppose that moral boundaries should coincide with the boundaries of our everyday community: not least because these latter boundaries are themselves not

> closed, but rather are defined in part by an increasing set of exchanges with distant strangers. (Corbridge 1993a: 463)

Current disciplinary preoccupations with porous places, different spaces, and dissolving the distinction between global and local (as 'glocal') underline growing recognition of changing geographical experience.

An example is provided by the way in which modern telecommunication can create a sense of moral engagement in distant lives. This was revealed in the occupation of Tiananmen Square in the Chinese capital of Beijing in 1989, which involved the use of symbolic monumental space to draw wider attention to political dissent: 'to open China up to the gaze of distant bystanders' (Adams 1996: 425). Empathising with peoples whose suffering is seen on television can motivate practical care, as in such extravaganzas as Children in Need (Silk 1998). But it might also induce numbness, indifference or worse: the 'trap' of aerial distance, enabling horrors associated with poor living conditions, for example, even to be made picturesque (Martin 1996: 588–9).

A particularly problematic case is that of 'virtual communities' in 'cyberspace', where a wide range of electronic social intercourse is possible (Kitchen 1998: 396–7). For example:

> People in virtual communities use words on screen to exchange pleasantries and argue, engage in intellectual discourse, conduct commerce, exchange knowledge, share emotional support, make plans, brainstorm, gossip, feud, fall in love, find friends and lose them, play games, flirt, create a little high art and a lot of idle talk. (Rheingold 1993: 3)

In short, people engage possibly far distant others instantly, in ways usually associated with close proximity, but without face-to-face interaction or bodily contact. Such networks are capable of developing their own moral codes based on reciprocity: 'a kind of gift economy in which people do things for one another out of a spirit of building something between them, rather than a spreadsheet-calculated quid pro quo' (Rheingold 1993: 59). However, these communities are a luxury of the well-off, whose scope for caring relationships via the internet is confined to others with similar resources. They may therefore reinforce relative privilege. Furthermore, there are some relationships of mutuality and care which actually require knowledge of the other as physically embodied, capable of touch, which cannot be regulated by the choice to switch on the computer. Remote communication may also promote evil, such as sexual exploitation and violence, by facilitating interaction among people who enjoy such things, and even encouraging behaviour which face-to-face relations might constrain (an issue to which we will return).

All this raises the question of how ordinary people might develop a capacity to transcend conventional identification with local and particular others, to extend their scope for care. Rorty (1989: 196) exhorts us to expand our sense of 'us' as far as we can, including ever more distant and different others, looking for marginalised people and trying to notice our similarities with them. Feminist writings on the ethic of care stress a relational rather than autonomous conception of the self, formed locally with close others, but which can lead to the spatial extension of concern: 'we learn to care for distant others by first developing close relationships to nearby others, and then recognizing the similarities between close and distant others' (Clement 1996: 85). We learn what it is like for children close to home to starve, and then recognise that distant children are like those close to home. An obvious limitation is that some people in some places are protected from the direct experience of children starving, and from other misfortunes from which poor people elsewhere suffer. Even an expanded sense of 'us' may not include all.

Recognition of the spatial extent of cause and effect relationships, as suggested by Corbridge (above), can be taken further. Obligations to care for family and friends arise from the fact that they are particularly vulnerable to what we do, but, as many other people are similarly vulnerable, the ethic of care has implications beyond our sphere of personal relationships. Thus:

> Those closer to us will *tend* to be more vulnerable to our actions and choices than those distant from us, and thus we are not obliged to weigh everyone's interests exactly equally. Yet insofar as those distant from us *are* particularly vulnerable to our actions and choices, we have special obligations to care for them. And to that extent, the conventional boundaries of the ethic of care break down. (Clement 1996: 73)

Against feminists for whom care is necessarily local, and others who oppose the notion of global moral concern, is an argument that feminist partiality actually implies global extension:

> This is the significance of all the work that feminists put into theorizing 'difference' and into trying to incorporate a diversity of racial and class consciousness into feminist theory. Cross-cultural connections, theoretical and practical, are highly revered feminist achievements. Thus, for feminists, 'global moral concern' does not call for exactly equal consideration for the interests of all persons, but it does call for substantially more concern for distant or different people than is recommended by nonfeminist partialists. (Friedman 1993: 86–7)

This concern arises not only from experience mediated by social and spatial relationships but also from the extent to which feminism esteems caring in itself, as a moral value.

The societal and personal mesh with the universal and particular, in an expanded ethic of care:

> To meet one's caring responsibilities has both universal and particular components. On the one hand, it requires a determination of what caring responsibilities are, in general. On the other hand, it requires a focus upon the particular kinds of responsibilities and burdens that we might assume because of who, and where, we are situated. (Tronto 1993: 137)

Her attention to 'the particular location in which people find themselves in various processes of care' underlines the importance of context, of the geographical situation.

Thus, against the evidence of parochialism in the actual practice of care is the possibility of its more extensive expression. 'Human beings can and do care – and are capable of caring far more than most do at present – about the suffering of children quite distant from them, about the prospects for future generations, and about the well-being of the globe' (Held 1993: 53). The challenge is to give substance to such sentiments, in the form of feasible political practice. The provision of spatially extensive care has to be institutionalised, not only for effectiveness but also because a right to care, as with positive welfare rights in general, cannot convincingly be linked to individual duty (unlike negative liberty rights which entail individual duties of non-interference). Benevolence as merely the inclination to care is not enough.

It remains in this section to note some arguments for universal beneficence emanating from mainstream moral philosophy, rather than from inheritors of Carol Gilligan's ethic of care. In calling for an alteration in the way people in affluent countries react to such situations as famine in poorer parts of the world, Peter Singer (1972: 23) elaborated the following principle: 'if it is in our power to prevent something bad from happening, without thereby sacrificing anything of comparable moral importance, we ought, morally, to do it'. This principle takes no account of proximity or distance:

> The fact that a person is physically near to us, so that we have personal contact with him, may make it more likely that we *shall* assist him, but this does not show that we *ought* to help him rather than another who happens to be further away. If we accept any principle of impartiality, universalizability, equality, or whatever, we cannot discriminate against someone merely because he is far away from us.
> (P. Singer 1972: 24)

He concludes that we ought to give as much as possible to famine relief and so on, perhaps to the point of marginal utility at which by giving more we would cause ourselves more suffering than we would prevent. He still argues along these lines, but the perspective appears to have shifted from utilitarianism to the mutual recognition of similarity of the human experience of pleasure and suffering which grounds some other contemporary univeralist accounts (P. Singer 1995: 222).

Another influential contribution is that of James Sterba (1981). He points out that it is only recently that philosophers have begun to discuss the question of what rights distant peoples might claim, stressing the practical point that, for them to have such rights, 'we must be capable of acting across the distance that separates us' (Sterba 1998: 57). Of the various moral grounds justifying the welfare rights of distant peoples, he stresses those which appeal to a right to life or a right to fair treatment. He argues that the satisfaction of needs basic to the life of some distant persons (from the perspective of better-off parts of the world) is unfairly restricted by the property rights of those with possessions surplus to what is required for their needs (see Chapter 7). This echoes P. Singer (1972: 29): 'From the moral point of view, the prevention of the starvation of millions of people outside our society must be considered at least as pressing as the upholding of property norms within our society.' The right to fairness involves the proposition that, from a position of disinterest, people would endorse limitations to a right to accumulate goods and resources, so as to guarantee 'a minimum sufficient to provide each person with the goods and resources necessary to satisfy his or her basic needs' (Sterba 1981: 104).

One of the strengths of Sterba's approach is its recognition of structural impediments to meeting the needs of the (distant) poor, in the form of the property rights which under capitalism protect the accumulated wealth of others. By comparison, literature on the ethic of care has little to say on such matters, beyond bemoaning the marginalisation of the values and practices of caring, nurturance, compassion and so on. However, Tronto (1993: 175) does say that an emphasis on care 'is probably ultimately anti-capitalistic because it posits meeting needs for care, rather than the pursuit of profit, as the highest social goal'.

Insofar as postmodern thinking is relevant to such a grand question as the spatial scope of care, it is likely to encourage parochialism. The stress on difference and particularity, while drawing attention to the specific needs of various groups of hitherto marginalised 'others', dilutes the force of an argument from human sameness or similarity, which supports spatially extensive responsibility for people who are like ourselves in morally significant respects. Postmodernism shares with conservative communitarianism the risk of undermining grounds for critique of parochial selfishness, playing into the hands of those who seek not so much to defend group or local privilege as to avoid having to defend it. Thus, the moral relativism (or nihilism) encouraged in some postmodern thinking is far from politically benign; instead of being merely an intellectual indulgence of the well-to-do, this perspective helps to entrench their privilege.

COMBINING THE ETHICS OF CARE AND JUSTICE

[W]hile care is essential to a morally adequate life and society, so too is justice.
(Clement 1996: 116)

Reconciliation of the conflicting claims of partiality as concern confined to particular close persons, and impartiality in the sense of universalising some conception of beneficence, requires harmonisation of the ethics of justice and care (Baier 1987). In the conventional account, justice refers to the supposedly masculine practice of approaching moral issues equipped with general principles and rules, compared with the supposedly feminist perspective of prioritising contextual decision-making and maintaining relationships of care. The spatial convention is that justice is for the public realm, care for the private. These stark distinctions have already been substantially eroded in debates following the introduction of the ethic of care. What is now at issue is to collapse a dualism which might otherwise force an unnecessary choice between care as conservative parochialism and justice as universalism indifferent to actual human relations.

First to insight provided by Marilyn Friedman (1991b; see also 1993). She recognises the moral strength of partiality, required by intimacy and close relationships, and essential to integrity and the good life, seeing 'sheer intrinsic value in the very benefiting of friends and loved ones' (Friedman 1991b: 818). However, she has two serious reservations. The first concerns the actual context: the moral value of partiality depends on the actual relationships it helps to sustain, which can be supportive but which could be abusive or oppressive – as in the case of a paedophile ring or the Ku Klux Klan, for example. Hence, partiality is morally required in a relationship to the extent that it contributes to protection of the vulnerable.

The second reservation concerns the unequal distribution of means of favouring loved ones:

> The one who really needs general moral attention is the person who lacks resources and who would not be adequately cared for even if all her friends and family were as partial to her as they could be because they, too, lack resources. There are systematic social inequalities among different 'neighborhoods' in the distribution of the resources for loving and caring. (Friedman 1991b: 828)

These inequalities matter to the partialist stance:

> [W]hether or not, and to what extent, someone benefits from certain partialist relationship conventions has a lot to do with her 'social location', the sort of luck she had in being born to, adopted by, or linked by marriage to, relations with adequate resources for caretaking, nurturing, and protecting. (Friedman 1991b: 829)

And from the perspective of the giver, she points out that, untempered by any redistribution of wealth or resources, partiality is the moral prerogative and responsibility of only those who are able to care. Thus:

> [T]here ought to be a distribution of resources for caring, nurturing, and otherwise favoring loved ones which permits as many of us as possible to do so . . . by viewing partiality as morally valuable because of what it ultimately contributes to integrity and human fulfilment, and by considering the reality of unequally distributed resources, we are led to a notion that sounds suspiciously like the requirement of moral impartiality. (Friedman 1991: 830–1)

She concludes that partialists might differ from impartialists in placing too little emphasis on global moral concerns.

The crucial geographical point is that a person's spatial location is very likely to reflect good or bad fortune similar to the accident of birth (see Chapter 7). Spatial inequalities in capacity to care (or disparities between capacity and need) will thus tend to perpetuate patterns of uneven development which are morally indefensible unless persons deserve the luck of being in a particular place with particular resources, and in a particular network of (more or less) caring relationships. Combating these inequalities or disparities is a matter of justice.

An attempt at reconciliation from the more conventional perspective of impartiality is provided by Brian Barry (1995). He considers that the 'battle' between impartialists and non-impartialists is bogus, because they are talking about different things :

> What supporters of impartiality are defending is second-order impartiality. Impartiality is here seen as a test to be applied to the moral and legal rules of a society: one which asks about their acceptability among free and equal people. The critics are talking about first-order impartiality – impartiality as a maxim of behaviour in everyday life. (Barry 1995: 194)

He claims that universal first-order impartiality is rarely advocated; as Marilyn Friedman (1991b: 818) points out, few moral philosophers these days would oppose partiality in everyday life. Barry goes on:

> All of us have only a finite amount of time, attention, care, and affection to devote to other people (or to ourselves for that matter), and life would scarcely be worth living if we could not decide for ourselves – once we had met our general social obligations – on whom these should be bestowed. (Barry 1995: 201)

The caveat is vital: general social obligations come first, within a set of institutions consistent with second-order impartiality. Thus:

> What is required is a set of rules of justice . . . that provide everybody with a fair opportunity of living a good life, whatever their conception of the good may be, while leaving room for the kind of discretion in shaping one's life that is an essential constituent in every conception of the good life. (Barry 1995: 206–7)

Hence impartial justice and Gilligan-inspired care can be reconciled: they function at different orders or levels of moral deliberation.

Another attempt is provided by Onora O'Neill (1996), who argues against the separation of justice (with its universalist aspirations) and virtue (with its particularist tendencies). She constructs a general principle of justice in the style of Kantian universalisation, based on the prevention of injury. She approaches the virtues with a similar argument, leading to the conclusion that, like injury, indifference to and neglect of others cannot be universalised (we all depend on not being treated so), therefore some form of care and concern follows. But it is bound to be selective:

> Although many ethical traditions extol universal benevolence, love for all mankind, or concern for all, their rhetoric misleads. Justice can be observed in relations with all others; social virtues cannot . . . The social virtues . . . do not require generalized or maximal benevolence or beneficence . . . but only *selective* and *feasible* help, care, love, generosity, support or solidarity. (O'Neill 1996: 195)

There are compelling reasons for directing much care and concern to those who have become dear or near, for they have come to expect it. There can be no universal obligation, to which rights to care can be linked, only choice and opportunity. However, solidarity and rescue may offer more universal scope:

> *Solidarity* can be expressed across large spatial and social distances, to others who are neither near or dear, through forms of help and support for distant strangers, especially those who are destitute or oppressed . . . Acts of *rescue* are more dramatic expressions of care and concern, directed towards others in present danger and misfortune, who may also be neither near nor dear. (O'Neill 1996: 197)

Examples include charities and disaster relief. Such institutionalised activities differ from those kinds of care which seem intrinsically local. However, rescue can also be an individual act raising complex moral issues, as we will see at the end of this chapter.

There are substantial differences between the philosophical approaches summarised above. Friedman espouses a feminism critically aware of the parochial limitations of partiality, Barry is committed to impartiality embedded in liberal perspectives on justice, while O'Neill combines Kantian universalism with virtue ethics. But what is more interesting is the common ground. All three find impartiality as justice and partiality as care compatible, if in somewhat different ways.

Similar conclusions can be found in the work of others. For example, Dancy (1993: 178) combines the demands of benevolence as a neutral (impartial) virtue and the flexibility of the agent–relative (partial) virtue of care for one's family and friends: 'These virtues are in tension, perhaps, in terms of the demands they place on action. But they are not contradictory . . . It is possible for the same person to have both virtues at once.' Miller (1992: 200–1) echoes Barry in referring to different spheres, with partiality towards family practised against background institutions directed towards improving the lot of the disadvantaged. So do Mulhall and Swift (1996: 293), in an attempt to reconcile aspects of liberalism and communitarianism: once the demands of justice have been met, we are normally entitled to use our resources partially if we wish. To Benhabib (1992: 180), Gilligan's work 'does not provide us with sufficient reasons to want to reject universalist moral philosophies', but is more a contribution to contextual sensitivity. And for Geras (1995: 98), with respect to such solidarity as might reach people who are many and distant, 'the necessity of a generalizing moral rationality to work together with decent human sentiments would seem to be elementary': hence a combination of universalising notions and empathy is required.

Tronto (1993: 171) gets to the heart of the matter: 'care needs to be connected to a theory of justice'. For Mendus (1993: 25): 'unsupported by considerations of justice and equality, care may simply not extend reliably beyond the immediacy of one's own family, or group, or clan, to the wider world of unknown others.' For V. Held (1993: 76), an exclusive focus on care may lead us to accept a particular economic stratification in which the rich care for the rich and the poor for the poor. And for Solomon (1995: 259–60): 'Without emotion, without caring, a theory of justice is just another numbers game . . . Reason and emotion are not two conflicting and antagonistic aspects of the soul. *Together* they provide justice, which is neither dispassionate nor "merely emotional"'.

Grace Clement (1996) provides the most thorough account of what it means to integrate the care and justice perspectives. She argues that both are involved in how people live and evaluate: both can shed different light on situations, each is required for the other, both are distinctive as well as interdependent. The ethic of justice requires contextual details as well as abstract principles, while the ethic of care requires general principles as well as contextual details. For example a focus on care could provoke questions about the justice of treatment of care workers, while a focus on justice could raise questions about the distribution of resources for care. Thus, care and justice should not be seen as competitors, but as allies. As to the spatial scope of beneficence, far from undermining our relations of justice to distant people, a care perspective helps us to recognise our justice obligations to them:

> [I]f we want to take these obligations seriously, we should try to extend our sense of social connection by eliminating the barriers, such as ignorance, that make us feel disconnected from those physically and culturally distant from us.
>
> (Clement 1996: 85)

This observation gets to the core of a problem on which the aspirations of global concern may yet founder. It is a problem of metaethics: of self-identity and moral motivation. If, as has been suggested by various writers, extending the spatial scope of beneficence depends on the capacity to empathise with distant others whom we understand to be similar to ourselves, this requires something more than the kind of relational (or social) self found in feminism and communitarianism, imbued with a special sentiment capable of motivating care for nearest and dearest but which may remain parochially confined. Extending empathetic capacities developed in relationships with closely known persons to those more distant does not come solely from motives rooted in self-identity: it also requires reasoning and analogical insight:

> [G]lobal moral concern is a rational achievement but not an immediate motivation. It is, furthermore, an achievement only for some selves. It is a result of moral thinking that has no necessary motivational source in the self, so not everyone will find it convincing . . . it brings us face to face with what seems to be one important limit of that concept: its inability to ground the widest sort of concern for others in unmediated constituents of the self. We, thus, confront the apparent fragility of the human motivation of global concern, even in socially constituted selves.
>
> (Friedman 1993: 88)

To achieve the grounding required, and the moral progress implied, may be no less a task than the completion of a transformation of human identity initiated when people were first inclined to look beyond their own highly localised experience, to see others with self-recognition and sympathy.

THE CONTEXTUAL EXPERIENCE OF MORAL LEARNING

> [P]eople cannot be just or moral in a vacuum. (Lafollette 1991: 331)

This brings us to the question of how people learn morality, to do right and to be good (or not), and of the significance of the context in which this takes place. What kind of society, in the sense of its social relations, spatial associations and political economy, is conducive to the development of the morality required to underpin a practical concern for the well-being of others? If it is right and good to care for others and especially those in greatest need, and to avoid harming others and especially those most vulnerable, how might such sentiments be induced?

Two approaches to moral development compete in contemporary communitarianism. One stresses the inculcation of social norms by conditioning or socialisation, the other their discovery and reconstruction in the experience of social interaction. The first may be exempified by Etzioni (1995: 95–9), for whom successful socialisation involves 'character formation', entailing the acquisition of personal traits conducive to moral conduct, and the internalisation of commitments to a set of substantive values through 'moral education'. Persons are more than passive containers of received wisdom, but the scope for critique is limited:

> We gain our initial moral commitments as new members of a community into which we are born. Later, as we mature, we hone our individualized versions out of the social values that have been transmitted to us. As a rule, though, these are variations on community-formed themes . . . we find reinforcement for our moral inclinations and provide reinforcement to our fellow human beings, through the community.
>
> (Etzioni 1995: 30–1)

Etzioni's communitarianism has been criticised as concealing the distinction between functional necessity and moral autonomy (Bowring 1997: 112); it assists the reproduction of a particular kind of society, of parochial conservatism lauding the values of nation and family, rather than encouraging more extensive other-regarding behaviour.

The perspective of moral development through social interaction is exemplified by Selznick (1992). He points out that 'the minimally or perversely socialized person may be wounded or brutalized, spiritually impoverished, socially incompetent . . . socialization is a precarious venture' (Selznick 1992: 124). He asserts that 'authentic human ideals have material foundations. They are rooted in existential needs and strivings . . . genuine values emerge from experience; they are discovered, not imposed' (Selznick 1992: 19). Thus, social interaction is the 'forcing house' of moral development as the capacity to overcome parochial perspectives and adopt the standpoint of ever larger communities and universal values. There are echoes here of Sack's *Homo Geographicus* (1997), outlined in Chapter 1.

If moral development is experiential and relational, the context in which it takes place must be important:

> Without appropriate social support, moral development is frustrated or distorted; and moral recalcitrance is fostered in some social settings, restrained in others. Therefore the most important aspects of the human condition, viewed in the light of moral experience, are the nature and quality of social participation.
>
> (Selznick 1992: 183)

Being other-regarding as well as self-preserving is strengthened by 'organic ties to persons, history, deeds, and nature' (Selznick 1992: 184); in other words, by ties of geography.

The role of interaction within a community of sorts is quite widely regarded as crucial to moral development: 'The human infant becomes a "self", a being capable of speech and action, only by learning to interact in a human community' (Benhabib 1992: 5). The importance of spatial scale has been explained as follows:

> [M]oral capacity is fostered, cultivated, and exercised within the social environment of the small-scale setting – or it is *not* acquired at all . . . what is fostered here, in the setting of proximity, is the capacity for developing *empathy* with others . . . the

> faculty that *underlies* and so facilitates the entire series of specific, manifest emotional attitudes and ties to others, such as love, sympathy, compassion, or care.
> (Vetlesen 1993: 382)

It is important to stress the dimension of gender, of mothers as principle carers of children and of women's traditional role as moral educators (Clement 1996: 110), in the early and necessarily localised process of moral learning.

An issue in moral development crucial to the spatial extension of benevolence is how the setting might affect the way people treat strangers, beyond the intimacy binding nearest and dearest. Interacting regularly with particular local others, experiencing mutuality and reciprocity, might encourage wider application of such sentiments:

> Close personal relationships are grist to the moral mill . . . We can develop neither the close moral knowledge nor the empathy crucial for an impartial morality unless we have been in intimate relationships. Someone reared by uncaring parents, who never established close personal ties with others, will simply not know how to look after or promote the interests of either intimates or strangers. (Lafollette 1991: 330)

However, the additional demand posed by strangeness may require special moral effort, even rules. Premodern communities often incorporated a formal ethic of hospitality towards non-hostile others. Thus, the extended family of traditional African society, for example, is capable of including anyone: 'no-one is a stranger' (Shutte 1993: 50).

The existence of a certain sort of community attitude might also help to explain not only beneficence but also how people can behave badly towards others. Sterba (1996: 442–3) speculates that in communities characterised by a *laissez-faire* morality, according to which one's basic duty to strangers is not to harm them, evil is possible because morality is insufficiently concerned about the welfare of others, whereas communities which accept a strong duty to help strangers will also prevent evil. Something similar might be said about attitudes to those within the community whose conspicuous difference makes for uneasy membership: hostile rejection could provoke a dangerous reaction that sympathetic incorporation might avoid.

If empathy is important to the treatment of different others, social relations may be a crucial consideration. Sypnowich (1993a: 494, 496) refers to the unfortunate combination of 'pursuit of self-interest in the market matched by exclusive altruism in the family', and to the accompanying inequalities that divide people, and concludes that 'injustice limits the extent to which empathy is possible'. The impartiality of the 'hidden hand' which is supposed to be a virtue of market relations also implies impersonality, or lack of attention to context. Those forces of contemporary capitalism which differentiate and divide make for a particularly vicious circle, constraining the extension of human solidarity and sympathy based on 'the imaginative ability to see strange people as fellow sufferers' (Rorty 1989: xvi), rather than merely as potential competitors or customers.

What of spatial scale: may distance alone dull moral sensibility? Since Aristotle there has been a recognition that the strength of such sentiments as compassion, pity and envy may be reduced with increasing distance from their objects (Ginzberg 1994). It may be easier to kill at a distance than face to face, facilitated by the adoption of modern methods of warfare which reduce the destruction of places and people to a technical exercise. For example Todorov (1999: 162) refers to evidence of a correlation between altitude and attitude during the Vietnam war, when pilots bombing from a great height had clear consciences while those firing from helicopters experienced remorse and anguish. Looking back to his wartime experience, former British Prime Minister Edward Heath alludes to being in charge of a firing squad and to the traumatic experience of watching a man dying at close quarters, concluding:

> It is one thing to be in the war when you see the enemy on the other side and you bombard them. Then later on when you pass over their ground you see the dead bodies lying around. But it is very different when you see an individual.
>
> (*Guardian*, 24 September 1998)

Bauman (1989: 151–68) and others cite the experiment of Milgram (1974), who showed that inhumanity is a matter of spatial as well as social relationships. He found an inverse ratio of readiness to cruelty and proximity to victims, by observing that the proportion of subjects in his experiments who would inflict pain on another increased from when it required physical contact to when it involved manipulating an instrument, and to when victims were hidden from view. Physical propinquity is therefore claimed to be morally significant.

In the end, a combination of social, spatial and personal considerations must be implicated in the propensity for evil, and for good. But if the context itself is disturbed and unstable, the constraints even on unexceptional people may be loosened, with both horrendous and heroic consequences.

A MORAL GEOGRAPHY OF GENOCIDE AND RESCUE: THE HOLOCAUST

As an extreme event in the history of human evil, the Holocaust is a familiar testing ground for moral theories. On the face of it, genocide seems incomprehensible, impossible to reconcile with any ethics. It is as if moral attitudes disappear when we need them most, in situations of upheaval, 'when the fabric of moral interactions which constitute everyday life are so destroyed that the moral obligation to think of the other as one whose perspective I must weigh equally alongside my own disappears from the conscience of individuals' (Benhabib 1992: 138). Yet even in these circumstances, some persons reveal unimaginable courage in aiding others at risk. The intention here is not even to summarise the vast Holocaust literature, but merely to sample it insofar as the role of distance and proximity are concerned.

We begin with Zygmunt Bauman (1989). He points out that those who perpetrated the Holocaust were not, for the most part, pathologically insane, like mass murderers, but ordinary people (see also Goldhagen 1996). Moral inhibitions tend to be eroded when the violence is authorised officially, routinised by rules and roles, and victims are dehumanised by definition and indoctrination; also when the victims are concealed, with the perpetrators distanced physically and/or psychologically from the consequences of their decisions (Bauman 1989: 21–5). He argues that responsibility arises out of the proximity of the other, and that neutralisation of the moral urge involves replacing proximity with physical or spiritual separation. It was this which made the Holocaust possible, facilitated by the technical and bureaucratic achievements of a modern rational society (Bauman 1989: 184). Thus, 'Eichmann's conception of duty neutralized the immorality of what was being done and distanced him from the experience of his victims', but his character was shaped by a distinctive milieu: 'In the right circumstances, where similar justifications and distancing mechanisms are available, almost anyone's conscience can be twisted and deformed' (Selznick 1992: 177–8).

Bauman (1989: 186) explains that, in those towns and cities where Jews formed a large segment of the population, relations between Germans and Jews were hardly hostile, even in the first few years of the Nazi era. Attempts to inflame anti-Jewish sentiments were not particularly successful, and those responsible for propaganda found it hard to get people to see the negative stereotype in the Jews they actually knew – the 'Jew next door'. Thus, genocide could not be accomplished until the Jews 'had been removed from the horizon of German daily life, cut off from the network of personal intercourse, transformed in practice into exemplars of a category' (Bauman 1989: 189). This was achieved by the spatial concentration and separation of the ghetto, as in Łódź (see Chapter 3). His general proposition is as follows:

> [M]orality seems to conform to the law of optical perspective. It looms large and thick close to the eye. With the growth of distance, responsibility for the other

shrivels, moral dimensions of the object blur, till both reach the vanishing point and disappear from view. (Bauman 1989: 192)

This interpretation has been challenged by Arne Vetlesen (1993). He argues that proximity interacts with a number of other factors, requiring a closer look at living with others which 'situates the human individual within a social, intersubjective context' (Vetlesen 1993: 381). It is within this social embeddedness that the individual develops personal identity. He distinguishes between small-scale and large-scale settings of interaction (corresponding with proximity/distantiation, and with face-to-face/indirect encounters), and suggests that it is these that makes a moral difference. Moral capacity is developed through living with others, becoming its subject by virtue of being its object; responsibility is not pre-social, but created. So:

[F]rom the concentration through the deportation to the destruction of the Jews, what the Nazi perpetrators took pains to achieve was the *removal* of the victims not merely 'from sight' but more basically, from the small-scale setting *per se* and so from the moral-emotional dynamics intrinsic to it. (Vetlesen 1993: 383)

Bauman (1993: 143–4, 185) acknowledged the force of Vetlesen's argument that responsibility arises from the experience of empathy.

Other debates have focused on the protection or rescue of Jews. Richard Rorty (1989: 189) suggests that the chances of a Jew being hidden by gentile neighbours was greater in Denmark and Italy than in Belgium: 'many Danes and Italians showed a sense of human solidarity which many Belgians lacked'. He asserts that, rather than saving Jews because they were fellow human beings, Danes and Italians would have explained their actions in more parochial terms, such as the particular Jew they were taking risks to protect was a fellow Milanese or Jutlander, a fellow member of the same union or profession, or some such association:

[O]ur sense of solidarity is strongest when those with whom solidarity is expressed are thought of as 'one of us', where 'us' means something smaller and more local than the human race. That is why 'because she is a human being' is a weak, unconvincing explanation of a generous action. (Rorty 1989: 191)

However, Norman Geras (1995) has responded by showing that people who rescued Jews were motivated by universalist sentiments, as well as by familiarity with specific Jews. Referring to one set of findings (Oliner and Oliner 1992):

[A] large majority of rescuers emphasized the ethical meaning for them of the help they gave: some of these in terms of the value of equity or fairness; more of them in terms of the value of care; but in any case with a sense of responsibility common amongst them that was 'broadly inclusive in character, extending to all human beings'. (Geras 1995: 22)

This claim is substantiated by repeated stories of individual rescuers (mainly from Block and Drucker 1992). Hence the conclusion that 'a universalist moral outlook appears to have had a very significant part in motivating Jewish rescue. Many rescuers give voice to it and few do not . . . no rescuer I have come across overtly repudiates it' (Geras 1995: 36). While physical proximity may have helped to induce the sympathy towards close others which constituted a necessary (if not sufficient) condition to rescue them, a significant number of people were able to transcend this kind of particularism, and to find moral value in so doing. The underlying philosophical issue is the tension between the contingent character of Rorty's notion of solidarity, and the universalism for which Geras finds evidence in actual human behaviour manifest in the most demanding circumstances.

Returning to the Polish experience, with which the previous two chapters concluded, it is worth asking what can be learned from this particularly important context. The Holocaust was carried out largely in Poland; during the German occupation three million Polish Jews died (almost 90 per cent of them all), along with about two million Poles (nearly 10 per cent). Writers on Polish-Jewish relations point to the strong and traditional anti-Semitism, to the large number of Jews and to their low level of assimilation (Tec 1989: 301), and to fear and hatred between the two communities which had begun to grow before the outbreak of war (Polansky 1989: 226). The Germans might therefore have chosen fertile ground for genocide, with limited resistance to be expected on the part of the non-Jewish population. Polish response to the Holocaust defies simple explanation, being 'conditioned by a tangle of political, social, and psychological factors'; however, 'present and past, reality and fantasy interacted to create immense barriers not just to helping Jews but to wanting to help them' (Steinlauf 1997: 30). Furthermore, Poles who aided Jews assumed risks largely unknown in Western Europe, including execution of entire families and circles of people (Lukas 1997: 143). Nevertheless, it has been estimated that a million or more Poles were involved in helping Jews (Polansky 1989: 240; Lukas 1997: 150). A quarter of the 'Righteous Gentiles' whose rescue of Jews is recognised at the Yad Vashem Holocaust Museum in Jerusalem are Polish.

It could be that different responses generated by national or local contexts were less significant than those arising from the diversity of individual human agency. The rescuers were obviously distinctive:

> While most people saw Jews as pariahs, rescuers saw them as human beings. This humanitarian response sprang from a core of firmly held inner values. These values, which included the acceptance of people who were different, were unwavering and immutable. (Fogelman 1994: 6).

Some investigations claim that those who saved Jews were characterised by social marginality, separateness or individuality, which enabled them to stand up for their beliefs (Tec 1986). Others found that the vast majority felt a sense of belonging to their community (Oliner and Oliner 1992; Fogelman 1994). As to the link between gender and style of moral reasoning, as revealed to researchers by rescuers:

> Both men and women showed morality based on caring and attachments . . . had emotional responses to the plight of Jews . . . came to the aid of victims through an outraged sense of justice . . . Carol Gilligan's different voice was not so different after all. It was a shared voice of common decency and humanity.
>
> (Fogelman 1994: 240, 251).

Such values, instilled in childhood, were associated with a nurturing and loving home, altruistic parents as role models, tolerance for people who were different, and experience of care and caring (Fogelman 1994: 254). In short, they arose in a context which could be created almost anywhere.

We conclude with the story of one individual's complex circumstances and motivations (Hoffman 1998: 207–8). It is of a woman who, like many others, was an inadvertent witness of terrible things, which by accident of proximity implicated her in life-and-death choices. Her decision to help rather than turn away appears to have been motivated by universalism rather than partiality: 'A human being is a human being . . . I was helping human beings'. Before the war she had a close Jewish friend, but hardly spoke favourably of Jews in general. Her own interpretation of her response, when asked for assistance, was in terms of people needing help, and of the message of her Catholic faith. But there was also the thought that someone else might one day help her brother. Perhaps, 'through her altruistic gestures she was propitiating fate, making sure she was not tipping the scale of justice, or of mercy, to the wrong side' (Hoffman 1998: 208). But perhaps she was also able to add to the chance of circumstance and the dictates of religion that imaginative capacity for universalisation, seeing her brother in the place of those she helped, which coupled a moral intuition of beneficence with a courage strong enough to risk her life.

CHAPTER 6

SPACE AND TERRITORY: WHO SHOULD BE WHERE

> The political map, the plethora of ethnic-historical legacies, the patchwork of societies, faiths, communal identifications across our globe teems with unresolved conflicts, with religious-racial enmities, with non-negotiable claims to an empowering past, to sacred grounds. (Steiner 1997: 48)

The literary scholar George Steiner prefaces some reflections on the Jewish people with observations which could have come from a geography textbook. He suggests that the Jewish condition embodies what modern physics calls a singularity: 'a construct or happening outside the norms, extraterritorial to probability and the findings of common reason'. In short, it is different. Against all the odds, the Jews have survived, in the distinctive geographical contexts of the Diaspora and the tightly constrained (and contested) territory of the State of Israel:

> After more than two millennia of systematic and fitful persecution, of scattering into exile, of suffocation in the ghetto, after the Holocaust Jews insist on existing *contra* the norm and logic of history, which, even barring genocide, are those of gradual melting, assimilation, cross-breeding and the effacement of original identity.
>
> (Steiner 1997: 49)

This struggle for identity is a common theme in fiction, for example in Isaac Bashevis Singer's novel *Shadows on the Hudson* (1998), which captures an eroding Jewishness among immigrants to America in the aftermath of the Holocaust. His central character Hertz Grein, a Talmud teacher turned financier and philanderer, finally returns to the ultra-orthodox neighbourhood of Mea Shearim in Jerusalem, to rediscover what it means to be a Jew.

Human identity is closely associated with territory, as a subdivision of space which a group of people have come to occupy and in some sense make their own. This chapter explores contesting moral claims to geographical space, exemplified no more vividly than in Israel. In so doing, we try to reveal something of both the singularity and generality of the question of who should be where. So many and diverse are the strands connecting geography with ethics in this context that what follows is highly selective. A review of practices of inclusion and exclusion leads to the morality of different ways in which claim to territory are made, illustrated by Israel/Palestine. A discussion of multiculturalism and minority rights sets the scene for a case study of local conflict over space, in the city of Jerusalem.

INCLUSION AND EXCLUSION

Tension between the sharing and division of space, with inevitable conflicts, is a continuing preoccupation of the discipline of geography (e.g. Chisholm and Smith 1990). The most obvious expression of the bounding of space is the contemporary nation state: 'territorial containers' covering almost every portion of the settled world (Taylor 1994). The nation state provides a geographical framework for sovereignty and for the rights and obligations of citizenship. The political significance of the congruence of nation, state and territory was one of the major revelations of the modern era. Deviations from this ideal generate some of the most bitter conflicts in the contemporary world, as in the defining case of the Balkans.

The relationship between territory, boundaries and identity has been responding to changes in context in recent years. These include the geopolitical disintegration of Eastern Europe, the resurgence of various forms of nationalism, the revival of ethno-regional separatist movements and the consolidation of transnational organisations. In the contemporary world of globalising economic and environmental relations, and of postmodern suspicion of fixed categories, the relevance of the conventional political map of sovereign nation states is increasingly being called into question (e.g. Taylor 1995; Newman and Paasi 1998).

An illustration of the critique is provided by Michael Shapiro, whose explanation of some implications of nation states for personal identity deploys the concept of moral geographies (see Chapter 3). He suggests that the dominant geopolitical map is 'a moral geography, a set of silent ethical assertions that preorganize explicit ethico-political discourses' (Shapiro 1994: 482). The construction of national stories legitimating state boundaries of inclusion and exclusion are a normalising strategy for the *status quo*, against which may be posited a view which stresses flows of people. The nation state is being challenged, in terms of control over territory, by subnational affiliations, so an ethics of postsovereignty must address changing identity spaces. The map of nation states, which continues to supply the 'moral geography that dominates what is ethically relevant', should be replaced by a more equitable map, incorporating an ethic of respect for difference (Shapiro 1994: 495, 499).

Here are glimpses of a perspective with postmodern undertones, incorporating the good of boundary opening, privileging motion and recognising the importance of difference and the plight of marginalised 'others' within conventional territorial state structures. There are echoes of these positions in contemporary political geography, acknowledging the possible demise of the nation state and its implications for the spatial organisation of governance (e.g. Ó Tuathail 1996; Taylor 1996). Alternative territorial arrangements have also attracted attention in ethics. For example, there is an argument for a multilayered organisation in which sovereignty is distributed 'vertically' (by geographical scale) as well as 'horizontally' (by location), so that a decentralised democracy could bring some decision-making close to the people affected, while a higher level would be required for issues with wide spatial scope, like economic and environmental justice (Pogge 1992).

The advocacy of political and administrative decentralisation to empower local authorities and communities is part of a broader contemporary focus on civil society or citizenship at the sub-national scale (S. Smith 1989). Civil society comprises households, family networks, civic and religious organisations and communities, 'bound to each other primarily by shared histories, collective memories and cultural norms of reciprocity' (Douglas and Friedmann 1998: 2).

This reference to community, and the sentiments associated with it, connects the practical issue of territorial organisation to central issues in moral and political philosophy associated with conflict between communitarianism and liberalism (outlined in Chapter 4). The nation state lends itself to such liberal projects as the extension of rights and redistributive possibilities over relatively large territories, to transcend inherited local advantage. But the separation of individuals from community dilutes the mutuality on which inclination to beneficence may depend (as explained in the previous chapter). Thus: 'Citizens are asked to sacrifice more and more in the name of justice, but they share less and less with those for whom they are making the sacrifices' (Kymlicka 1993: 375). However, local empowerment may merely mask a return to communitarian partiality, and preservation of privilege. This dilemma takes on special significance when, to the spatial scale of political organisation and decision-making, is added the issue of movement, and especially the possibility of persons crossing hitherto protected borders.

The unequal life opportunities arising from the accident of birth in a particular place, which pose a major problem for social justice (see next chapter), can in principle be overcome if movement is an accepted part of personal freedom. Indeed, moving around in search of a better life is among the individual initiatives applauded by liberals. It might even be viewed as a human right:

> The fact that we are born with legs and intelligence opens up to us ever new spatial and intellectual horizons. The human race, like other animals, is a migratory species . . . The human ability to migrate has been one of our basic assets of survival, allowing us to free ourselves of geographical constraints, from bondage to the earth. (Tucker 1994: 186)

A paper the title of which postulates free population movement as 'the civil right we are not ready for' elaborates:

> At some future point in world civilization, it may well be discovered that the right to free and open movement of people on the surface of the earth is fundamental to the structure of human opportunity and is therefore basic in the same sense as free religion, speech, and the franchise. (Nett 1971: 218)

Liberal theory 'has difficulty justifying boundary-keeping projects that disadvantage outsiders who want in, or insiders who want to get out' (Becker and Kymlicka 1995: 466–7). However, actual practice suggests that territorial boundaries are protected with increasing rigour, which is more in keeping with communitarian arguments for restrictions on entry so as to preserve collective identity.

Discussions of borders and movement in moral theory make a variety points critical of closure and exclusion. For example:

> Citizenship laws combine, in different ways, criteria of birth or descent . . . and territory . . . These ascriptive criteria are morally no more defensible than all the other, like kinship, sex, age, region, residence, language, habits, culture, lifestyles, gender, religion, nationhood, social class, membership in churches, parties, and so on. (Bader 1995: 214)

Further: 'Citizenship in Western liberal democracies is the modern equivalent of feudal privilege – an inherited status that greatly enhances one's life chances', and similarly hard to justify (Carens 1987: 252). The power to exclude aliens is inherent in sovereignty, giving every state the right to exercise that power in its own national interest; however, the libertarianism advocated by Nozick (1974), the liberal approach to justice set out by Rawls (1971) and even a utilitarian calculus lead to 'a position far more favourable to open immigration than the conventional moral view' (Carens 1987: 264). And to another, 'nothing could be more suspect, from the standpoint of justice as social freedom, than disadvantages maintained by coercively regulated borders' (Miller 1992), raising the question of whether the claim of sovereignty is anything more than a form of special pleading (Beitz 1991: 246).

Given all this, and the proposition that inequalities in life chances among states cannot be defended morally, there is a case for 'fairly open borders' as a matter of justice, as opposed to the closure of 'collective welfare chauvinism' (Bader 1995: 215). However, firm restriction on entry is likely to continue to be the predominant practice. For example, the prospects of some form of political integration in Europe appears not to be making the British people more welcoming to foreigners, even if they are asylum seekers rather than 'economic migrants'. Indeed, the prospect of Western Europe being flooded with migrants from the east, as the postsocialist experience becomes ever more repelling, suggests the consolidation of an inter-bloc border as impenetrable as the old Iron Curtain. While Western capital exercises its taken-for-granted freedom of movement to the east, or any other direction its owners chose, the same freedom is denied Eastern European labour eager to move west. (For more on ethical issues in migration of people and money, see Barry and Goodin 1992, and on immigration and social justice, see Black 1996.)

Borders are often accompanied by unambiguous symbols of exclusion. Barbed-wire fences are no more welcoming than signs saying 'Keep Out', 'Trespassers will be Prosecuted' or 'Whites Only'. However, the way in which some people are subjected to spatial exclusion while others are admitted is usually more subtle. The landscape may be forbidding, literally, to some kinds of people. For example, the 'Islamisation' of some townscapes in Bosnia since the war in the early 1990s signals who belongs where in this bitterly contested land. There are also various discourses of difference, whereby particular personal characteristics are subject to moral evaluation which makes exclusion appear legitimate, even part of the natural order.

David Sibley has linked boundary consciousness to an urge to purify social space by excluding certain 'others'. He points out:

> [T]here are many recent instances of collective action against groups who appear to threaten the perceived spatial and social homogeneity of localities, where the threat comes from difference in ethnicity, sexual orientation, disability, or life-style . . . Because the boundary of the group is the main definer of rights, separating members from strangers, it is essential to maintain the boundary, either by *expelling* polluting agencies . . . or by *excluding* threatening groups or individuals.
>
> (Sibley 1988: 410, 411)

He claims that the utopian vision has always been one of homogeneity, conformity, stratification and separation, as for example in urban planning. Exclusionary discourse draws particularly on colour, disease, animals, sexuality and nature, but 'they all come back to the idea of dirt as a signifier of imperfection and inferiority, the reference point being the white, often male, physically and mentally able person' (Sibley 1995: 14). As mixing carries the threat of contamination, 'spatial boundaries are in part moral boundaries. Spatial separation symbolizes moral order', with purification a key feature in the organisation of social space (Sibley 1995: 39, 77). The notion of ethnic cleansing is an obvious case in point, deployed by the Nazis half a century before the contemporary horrors of central Africa and former Yugoslavia.

There are links between this perspective and other contemporary work in cultural geography, on the placing of 'deviant' groups in the context of social control, and on their response in the form of transgression of imposed boundaries. Tim Cresswell (1996) argues that the place of an act is part of our understanding of what is good, just and appropriate. Transgressive acts are those judged to be 'out of place' by dominant institutions and actors, and may constitute a form of resistance; along with reactions to them, they underline prevailing values concerning what is 'in place'. Such acts are frequently described by their opponents in terms of disease and contagion. For example, hostile elements in the British press used images evoking dirt, smell, sexuality and such-like to portray women in their camp set up in the 1980s as a demonstration beside the American nuclear weapons base at Greenham Common, descriptions packed with signifiers of deviance directed towards readers' prejudices (Cresswell 1994).

The distinction between private space (usually taken to be that of the home) and the public arena beyond is of special relevance to the issue of exclusion and inclusion, including the gendering of these spaces and whether different codes of morality apply to them. Private space permits practices which might not withstand critical public scrutiny, and part of the struggle over what behaviour is right in particular places may involve (re)definition of private and public. For example, there are attempts to extend the environment of home, as a private retreat from the heterosexist (and often homophobic) public sphere, into so-called 'queer space' or 'gay villages' in some cities (Binnie 1995). A case in the other direction is the quasi-privatisation and surveillance of shopping space and the control of behaviour therein, in an attempt to create an attractive and secure spending environment, which can lead to 'the virtual disenfranchisement of young people whose appearance and conduct do not conform to the moral codes of well-ordered consumption' (Bianchini 1990: 5).

Lest it be thought that interest in spatial strategies of inclusion and exclusion is confined to geography, reference may be made to a study of exclusionary practice in the anthropology collection from which cases were drawn in Chapter 2 (Howell 1997). Nigel Rapport (1997: 89) describes local attitudes in part of rural northwest England ('Wanet') in terms of various everyday discourses which would morally exclude outsiders from the land ownership and use from which 'proper belonging in Wanet' is derived. He explains that, to some, being a local in Wanet entails an absolute exclusion of the outsider, whether foreign immigrant, National Park warden, tourist, or retired people from elsewhere. Not for Wanet locals the openness to unassimilated others celebrated by I. M. Young (1990b). The manner in which they deny the outsider access to 'local sentimental space' is described as 'a morality of locality' (Rapport 1997: 93).

Thus various practices, from the formal definition of political jurisdiction through the control of movement to local discourses of belonging, contribute to the creation of moral geographies prescribing who should be where. This leads to the general question of how entitlement to occupy territory is asserted as a moral claim.

CLAIMS TO TERRITORY: PROMISED LAND?

The human urge for territory is sometimes likened to the imperative attributed to animals (Ardrey 1971). But in contemporary geography the concept of territoriality is usually associated with the sense elaborated by Robert Sack (1986: 19): 'the attempt by an individual or group to affect, influence or control people, phenomena and relationships by delimiting and asserting control over a geographical area'. An example was provided by the company town, in Chapter 3. Territory is required to satisfy people's basic material needs, preferably in a secure setting conducive to the reproduction of individual and social life. Territory is also an important source of human bonding or identity, linking the present to the past as a fund of common memories or history, and to the future as common destiny. Furthermore, territoriality can nurture 'an ethic of care and concern for our fellow citizens and for the environment we share with them' (Friedmann 1992: 133).

These propositions emphasise the rich and varied role of territory in human affairs. Emotional as well as material attachment to particular territory contributes to human flourishing. Thus, 'the first place to which the inhabitants are entitled is surely the place where they and their families have lived and made a life' (Walzer 1983: 43). Some place in the world is itself a physical imperative: persons as individuals or groups occupy a portion of space, as a natural fact. It might therefore be argued that there should be human rights to place, such as the right of access to land required for home and community and the right not to have one's place taken away without due process of law and adequate replacement. A place to live was one of the rights included with 'all things necessary for life' by Thomas Hobbes in *The Elements of Law*, a right enforceable against the state which exists to protect it (Walzer 1983: 43). But such a right is to place in general, not to a particular place, and hence the question of what constitutes a defensible claim on a particular place or territory.

The answer is usually formulated at the level of the nation state:

> Every nation claims to stand in a specific and privileged relation to a homeland – the 'ground' in a near literal sense of its national identity. The nation provides its members with an inalienable collective property: the land in which they have the

> right to live their lives. It thus recovers a relation to the land – a spatial location and identity – which the modern world has destroyed for most people. (Poole 1991: 96)

While the internal distribution of the 'collective property' will depend on how individual or group claims are evaluated within the state's political economy and code of justice, it will be clear who has a claim and who has not.

However, there are circumstances where different nations lay claim to the same land. We will use one such case – the competition of Arabs and Jews for space in Israel/Palestine (following in part D. M. Smith 1994a: 264–70) – to reveal something of the generality and particularity of how claims to national territory may be imbued with moral credibility. One of the great geographical paradoxes of the twentieth century is that, in order to make a place for themselves in what they regarded as their historic homeland, the Jews displaced others who were thereby dealt a similar experience of exile. The legitimisation of immigration of the Jews after the Second World War has been interpreted as compensation for centuries of persecution in Europe culminating in the Holocaust, but transformed into claims elsewhere – in Palestine (Sayigh 1979: 56–78).

The State of Israel has a population approaching 5 millions, 17 per cent of whom are not Jews. This minority are Palestinian Arabs, citizens of Israel but deprived of some rights confined to people defined as Jewish. There are also about 1 million Palestinian Arabs living in the West Bank, and a further 700 000 or so in the Gaza Strip. These, together with the less densely populated Golan Heights, are the territories occupied by Israel in 1967, in which there are now about 180 000 Jewish settlers. In addition, there are well over 2 million Palestinian Arabs living largely in Jordan and Lebanon, many of them in refugee camps, having been displaced by the establishment of the State of Israel in 1948 and by subsequent extensions of Jewish settlement.

The traditional claim made by Jews to territory in Palestine is religious, and based on the Bible. Following a sympathetic source (Reagan 1996), God gave the land of Israel to Abraham and his descendants through Isaac and Jacob (*Genesis* 17: 8; 26: 2–5; 28: 1–4, 13–14), so the land belongs to the children of Israel or the Jews (*Romans* 1: 16). The Jews were warned that disobeying God would lead to exile, but that they would be restored to their land which they would reclaim from wilderness (*Deuteronomy* 28: 30). Jewish settlement in this region could therefore be portrayed as 'a religious imperative which no human law can negate' (Newman 1984: 147). However, God's gift was conditional: religious practices were required of the Jewish people, who were supposed to exercise stewardship of the land, all of which reinforced attachment to this particular territory (Houston 1978; Shilhav 1985).

To the religious case is added long-standing occupation by Jewish kingdoms, from about 1200 BC to AD 69. When conquest by the Romans sent the Jews into almost two millennia of exile, however, their relationship with the promised land changed. Not being in occupation prevented the Jews from performing those religious commandments which could be carried out only on this particular land. However,

> [T]hey continued observing rituals and customs which were directly connected to that country, such as festivals marking the changing of the seasons there, ceremonies relating to the specific plants which grow there, and most of all, in referring to specific places in the land as 'sanctified'. (Kimmerling 1983: 8)

Hence the peculiarity of religious association with land from which the Jews had subsequently been absent longer than their initial occupation.

These Jewish claims are countered by similar claims on the part of Palestinian Arabs. There is a connection between the religion of Islam and territory which generates a Moslem assertion of sanctity comparable with that of the Jews. Jerusalem in particular highlights the conflict: the Western or 'Wailing' Wall, so revered by religious Jews, is barely a stone's throw from the al-Aqsa Mosque and Dome of the Rock, from which the Prophet Muhammad is supposed to have risen to Heaven. The association between religion, territory and nationality thus serves a comparable function for both sides, who see the entire expanse of Palestine as having moral or religious value and use this as a means of legitimating their claim (Kimmerling 1983: 215–16).

There is also a long history of continuous Arab occupation of the land after the Jews had left, and perhaps even before they arrived. One historian explains this as follows:

> The Palestinians' claim is predicated on the right of ownership evidenced by uninterrupted possession and occupation since the dawn of recorded history. They lived in the country when the Hebrews (of whom the Jews claim descent) came and lived there for a comparatively short period. They continued to live there during the Hebrew (and Jewish) occupation. They remained there after the last Hebrew or Jew left the country nearly two thousand years ago.
>
> (F. C. Sakran, quoted in Kimmerling and Migdal 1994: xvi)

As in some other territorial conflicts, the claim to be 'first peoples', in the sense of having initially occupied unappropriated land, is part of the narrative of national history. And, as elsewhere, it is contested.

Claims to territory grounded in its religious importance are powerful. Indeed, 'claims that land of religious significance should be returned to its original owners may have an edge over claims for the return of lands whose significance for them is mainly material or economic' (Waldron 1992: 19–20). Religious rites may have to be performed in specific places, while similar land elsewhere can serve the same purely material purpose. But in the case of Palestine there are two competing claims on comparable grounds, and even if it could be shown that one was stronger than the other, for example that the Jewish sense of collective religious identity is much older than the discovery of Palestinian nationalism steeped in Islam, this is unlikely to convince those with the weaker case. Indeed, the very nature of a religious claim based on a god's will makes it unconvincing to those who do not believe in that particular deity. It is also by no means obvious that a religious claim is always (or ever) superior in a moral sense to all other kinds of claims. The mixing of labour, sweat and so on with particular land in fruitful occupation could be even more persuasive, especially to those who recognise no divine authority.

The religious case has therefore come to be supplemented by arguments of greater secular appeal. Indeed, some ultra-orthodox Jewish claims concerning the absolute sanctity of the Land of Israel hardly lend support to the presence of a secular society and state (Glinert and Shilhav 1991: 61). Hence the emphasis on historic occupation, or on the identity between people and territory without the necessity of a deity to bind them together. Dissociated from the religious practices on which the sanctity of the land depends, the Jewish case becomes one of the fulfilment of national aspirations manifest in simply having certain territory: 'territorial fetishism' or 'paganism where territory is the god' (Shilhav 1985: 120).

This kind of case is strengthened by working to improve the land. Going back to the land as a source of livelihood was an important ethic of Zionism, and once Jews began settling Palestine the struggle of the pioneers became enshrined in mythology:

> Overcoming the difficulties of conquering the land from nature, when nature included not only sand and swamp, but psychological, social, and political obstacles as well, gives the settler a deep emotional attachment to the land and a feeling of entitlement to it. Defining the struggle as primarily one against nature rather than against men or sociopolitical forces was necessary in order to give self-legitimacy to the Zionist activities. (Kimmerling 1983: 227)

The draining of the swamps in particular took on enormous symbolic significance. 'The swamps were not simply a hydrological phenomenon; they represented a series of dualisms: desolation versus rebirth; the ugly versus the beautiful; weakness and death versus heroism and life' (Bar-Gal 1991: 20). Similar sentiments are attached to David Ben-Gurion's vision of the greening of the Negev Desert. Insofar as the Palestinian Arab population entered the scene, they could be portrayed as not caring for the land, further justifying Jewish occupation.

Being on the land was very important, in terms of the law and customs of the region: 'not living on it or working it – and even more, the presence of another on it – put one's title at risk' (Kimmerling 1983: 20): hence the role of presence as a political means of controlling territory in Ottoman times. The Jews continued this practice, from early immigration to the contemporary strategy of settlement on the West Bank, referred to as establishing 'facts on the ground'. Arab presence, for example where public rights of access existed as remnants of nomadic society and where land was left fallow for a few years to improve it, may have been less individual and conspicuous than Jewish agriculture (Kimmerling 1983: 38, 107). Such practices were a source of conflict with settlers unable or unwilling to understand Arab land-use and associated rights.

However, there is no question about the close attachment which Arabs developed with their land, forged through generations of hard toil. The home settlement was central to people's sense of identity, community and place. Accounts of Palestinian life emphasise village and clan solidarity, forming 'a warm, strong, stable environment for the individual, a sense of rootedness and belonging' (Sayigh 1979: 10). There was an almost physical love of the land: 'where blood ties are the backbone of the social structure, the repetition of motifs of erotic geography is readily understood' (Rubinstein 1991: 29). Planting the potent olive tree as a symbol of land occupation is one of their current strategies of establishing facts on the ground.

While entitlement to territory in this part of the world has tended to be grounded in a combination of the material and spiritual, the changing geopolitical context is generating different kinds of claims, just as those of the past responded to the moral environment of the time. In pointing to the contradictions between Zionist nationalism and the absence of an idea of territorial sovereignty within Judaism, Yosseph Shilhav (1993) notes the contemporary pragmatism of ultra-orthodox Jews in participating in a secular political system from which they can obtain advantages. He suggests that sovereignty is no longer a dictation of norms, behavioural standards, cultural identity and so on, adding: 'it is essential to stop attributing moral significance – and even more so religious significance – to political sovereignty of the old type' (Shilhav 1996: 276; see also 1995).

In these new times, which in Israel may be referred to as 'post-Zionist' or 'post-ideological', three general conditions are involved in a process of refining or redefining claims to particular territory. The first is population movement. Israel's borders are open, to large numbers of Palestinian Arab workers commuting daily and to Jews from elsewhere exercising their 'right of return' which has brought hundreds of thousands from the former Soviet Union in recent years. Added to the increasingly secular character of Israeli society, this is diluting any sense of national identity based on Jewish religiosity or even cultural distinction (Waterman 1999). The second is safety or national security. This is tending to override the traditional religious-historic territorial vision under which any concessions in the form of 'land for peace' would have been inconceivable, and which will determine the details of who lives where as a Palestinian state emerges. The third is market power. Just as money from Jews in the Diaspora facilitated the purchase of land to extend Jewish occupation before the creation of the State of Israel, so today sites are being bought in Arab East Jerusalem and elsewhere, for prominent (and provocative) Jewish occupation.

The circumstances in which Jewish immigration began, consolidated and ended up with a new nation state in the old promised land were a triumph over the tragedy of past pogroms and attempted genocide. The Jews learned from awful experience to help themselves, and at a crucial stage in their history this meant, literally, to land occupied by others. Israel's far-flung responsibility to other Jews, reciprocated by support from international Jewry, is an unusually expansive if highly selective expression of the often parochial ethic of care. In stark contrast is the unequal and at times brutal treatment of nearby others, who happen not to be Jews, an inversion of the customary localisation of concern. Not to be a Jew is the difference which really matters. Fervent Jewish nationalism may be an understandable reaction to a history of persecution, just as Palestinian nationalism has been forged out of the recent experience of dispossession and domination. But history has also rendered impossible the coincidence of nation and territory here, without massive further population movements.

The moral geography of the age of 'postsoverignty', if this is indeed what is to come, will face no greater challenge than fulfilling what, within the conventional geopolitical context, would be construed as the rights of minorities in these troubled territories. And, given the increasingly fractured identity of the Jewish population, these minorities will include different Jews, as we will see in the final section of this chapter.

MULTICULTURALISM AND MINORITY RIGHTS

> Minorities and majorities increasingly clash over such issues as language rights, regional autonomy, political representation, educational curriculum, land claims, immigration and naturalization policy . . . Finding morally defensible and politically viable answers to these issues is the greatest challenge facing democracies today.
>
> (Kymlicka 1995: 1)

Introducing a book on multicultural citizenship and minority rights (Kymlicka 1995), Will Kymlicka points out that most countries today are culturally diverse. He cites an estimate that the world's 184 independent states contain over 600 living languages and 5000 ethnic groups. Examples of recent conflict generated by this expression of difference include the experience of ethnic Albanians in the Kosovo region of Serbia, and of the Kurdish people split between Iraq, Turkey and some other states. I shall briefly review the main issues raised by such expressions of human diversity in proximity.

Susan Moller Okin (1998) explains two meanings of multiculturalism. One refers to quests for recognition by groups including women, gays and lesbians, and some ethnic and racial minorities, as a politics of identity (or difference) on the part of those whose voices have hitherto been silent or subdued, but who do not usually claim their own culture. The other meaning is what is usually understood in the context of debates on group rights to defend a culture which provides its members with a way of life across a range of activities: social, educational, religious, recreational and economic. Unlike the first sense, the second is likely to involve a shared language and history, and to be spatially expressed with some sense of territoriality adding to other aspects of collective identity. While the distinction is not rigid, it does serve to highlight different moral issues.

The first sense may include what Kymlicka and Shapiro (1997) refer to as lifestyle groups (e.g. gays), advocacy groups (e.g. environmentalists) and identity groups (e.g. women, the disabled), though there may be overlaps between them. Such groups should pose no problem for liberal democracies, committed to individual freedom, mutual toleration and political pluralism. The fact that they do reflects the power of a pervasive morality which tends to marginalise certain lifestyles, personal characteristics and politics, in the interests of maintaining some supposedly 'mainstream' but usually conservative value consensus. Hence the political emphasis on making space (perhaps literally) for difference.

With respect to the second sense, Kymlicka (1995: 6) distinguishes between the situation in 'multinational' states, where cultural diversity arises from the incorporation of previously self-governing and territorially concentrated cultures into a larger state (e.g the former USSR), and 'polyethnic' states where cultural diversity arises from immigration (e.g. the USA and UK). Elsewhere, the term 'ethnocultural group' is adopted in the context of ethnicity and group rights (Kymlicka and Shapiro 1997).

The notion of minority or group rights is by no means uncontroversial. There are those who argue that upholding the individual rights fundamental to liberalism, such as freedom of speech and assembly, should be sufficient to guarantee unrestricted choice of a way of life. However, some important experiences may be enjoyed only as members of a collective, or community:

> Given the special moral significance that rights possess, it is difficult to see what could justify our confining that significance to things which matter only to individuals as separate individuals and our refusing to allow that the same sort of moral significance might attach to things of fundamental value which people can experience and enjoy only in association with others. (Jones 1994: 187)

Questions of group rights tend to arise when the freedom of a minority to engage in a distinctive way of life, perhaps involving their own language, cultural rituals and religious practices, is threatened by a majority with the power of legislation or other forms of coercion.

The importance of minority group rights is underlined by the historical experience of majorities seeking to create a homogeneous polity in the form of a united nation state. This has included population expulsion and compulsory assimilation, and even genocide. The different sorts of group rights recognised by Kymlicka (1995) distinguish in particular between self-government rights (e.g. the delegation of powers to national minorities, often through some form of federalism), polyethnic rights (financial support and legal protection for practices of particular ethnic or religious groups) and special representation rights (guaranteed seats within institutions of the larger state). These all have a spatial dimension: the local and regional structure of political devolution, the location of state supported institutions and the territorial basis of group representation.

The spatial disposition of minorities reveals considerable diversity: some are dispersed, but they often live in geographical concentration, with attachment to particular territory. Some have a number of concentrations, like African-Carribeans and Muslims in British cities. Some are confined to an enclave within a single nation state (e.g. the Basques in Spain), some display irredentism with people living in one nation state claiming to be part of another (e.g. the Armenians of Nagorno-Karabakh in Azerbaijan), while others straddle state borders (e.g. the Kurds). Some situations are more complex, with assertions of group rights testing national unity, South Africa being a case in point (see Chapter 8). The origins of the minority in question also vary: some may be aboriginal or 'first people', others refugees or migrant workers, yet others the remnants of earlier invaders or ex-colonial peoples attracted to the land of the coloniser. Understanding the geography as well as history of the particular situation is indispensable to a grasp of the moral issues involved, and to the design of morally defensible as well as politically feasible resolutions.

One of the most difficult philosophical as well as practical questions in multiculturalism is that of the acceptable limits of difference. Specifically, how should a nation state respond to demands to protect or sustain within it a culture or way of life which is itself intolerant or otherwise morally objectionable to the wider society? Communitarians might be expected to support diversity, creating 'protected spaces of many different sorts matched to the needs of the different tribes' (Walzer 1994: 78). Like their postmodern colleagues, they are suspicious of any overarching value system against which particular local practices might be judged beyond the pale. The position of liberalism is more ambivalent, as some recent debates reveal (e.g. Kymlicka 1995: 152–72; Kukathas 1997; Okin 1998). Positions range from active support of distinctive cultures even if they are internally intolerant, to non-intervention in such groups, to confining minority rights to groups that are themselves liberal.

While liberalism in principle leaves the good life to individual choice, multiculturalism raises normative issues which are hard to avoid unless 'anything goes'. This is one context within which the problem of normative ethical relativism, as the doctrine that what actually is right or wrong differs among cultures, confronts universalism as the doctrine that there are ways of comparing different cultures on the basis of better or worse (introduced in Chapter 2). While committed to an overarching conception of the good of individual freedom, liberalism will otherwise tend towards relativism in the sense of neutrality among alternative ways of life. However, this could lead to support (or at least acceptance) of cultures which are not only intolerant but damaging to their members, and from which exit is rigorously controlled.

Claiming that most cultures have as one of their principle aims the control of women by men, Okin (1998) stresses the tension between group rights and what actually happens within cultures, especially in the private sphere, where gender relations disadvantage women. She also expresses concern over the contemporary deference to difference: 'focusing only on differences among women and bending over backward out of respect for cultural diversity does great disservice to many women and girls throughout the world' (Okin 1998: 666). Exposing those local practices whereby women are dominated and oppressed may facilitate normative comparisons among cultures, on the basis of at least this dimension of better or worse. Such context-sensitive research may also reveal that it is not only women whose freedom is severely constrained and that internal intolerance can be externalised. One example, with the added ingredient of territorial competition among different and irreconcilable ways of life, provides a concluding case study.

CONTESTING LOCAL SPACE: WHOSE JERUSALEM?

Jerusalem, Friday. The sun has set and the Sabbath has begun. The crowd thickens in front of the Western Wall, sacred place of Judaism, men and boys in the large enclosure, women and girls confined to the smaller part of this gender-segregated space. Floodlights illuminate the scene. In the right hand corner of the men's enclosure, a group are gathered around a table covered by books. They are singing, some songs sombre, some more joyful, some accompanied by dancing – hand in hand, around the table. They are dressed smartly but informally, kippas on heads, prayer shawls just visible beneath some of the jackets: observant Jews, but not ultra-orthodox. They are led by an ebullient figure, Moses-like with a large beard. A few other men approach the group. They look different: dark suits, white shirts, black hats, with hair protruding in sidelocks: the appearance of the Haredi or ultra-orthodox Jews. Their demeanour signals distaste for the singers. One of them holds a book, and reads some edict, perhaps disapproving of the lack of solemnity. Some boys, small-scale replicas of the adult Haredi, are led to the scene. They are lectured to by one of the men: 'This is sin', he could be saying, 'so take note, and keep strictly to religious law which forbids such frivolity'. The singers respond to the intimidation with added exuberance, urged on by Moses. The songs get louder, the dancing quicker, and the Heradi eventually move away. A ripple of applause from some who watched the performance marks the end of this local contest of identity in place: of what an observant Jew may do, and where.

Jerusalem is a deeply divided and vigorously contested city. Until the 1967 war, it was split between Israel and Jordan, with the eastern part Arab and the western part Jewish. Since the Israeli occupation, a Jewish population has been established in East Jerusalem, now exceeding the 160 000 Arabs. Further substantial Jewish settlements are planned by the municipal authority, while an American millionaire helps finance the purchase of existing Arab property for religious Jews to extend their control. And beyond, in the West Bank, huge suburban extrusions create more of the facts on the ground which try to seal off East Jerusalem from incorporation into a future Palestinian state. Such is the tough reality of this territorial struggle over whose Jerusalem it is to be.

But there is another struggle underway. This is less publicised outside Israel than that between Jews and Arabs, but very much part of contemporary life, subject of local research (e.g. Hasson 1996; Shilhav 1983, 1998) and featured in a text on multicultural cities (Sandercock 1998: 177–82). This contest is between Jews with different religious orientation, most evidently the secular and moderately religious Jews on the one hand, the ultra-orthodox or Haredi (plural: Haredim) on the other. The struggle is not only cultural but also deeply philosophical: between the liberalism to which the modern State of Israel might be expected to be committed and a religious-fundamental version of communitarianism, between an ethic of toleration of difference and a belief system illiberal enough to test minority rights to the limits. It is a struggle being played out politically, at both city and national level. It is being played out demographically, as the rapid natural increase of the ultra-orthodox promises to raise their share of Jerusalem's Jewish population from the present 30 per cent to 40 per cent in the next ten years. And it is being played out geographically, in residential and public space, as the Haredim seek to extend their occupation and control in a classic exercise of territoriality.

The ultra-orthodox enclave which now covers much of the northern part of West Jerusalem began in the late nineteenth century, with the establishment of neighbourhoods such as Mea Shearim and Batei Ungarin north-east of the Old City (Figure 6.1). Settlement has subsequently been

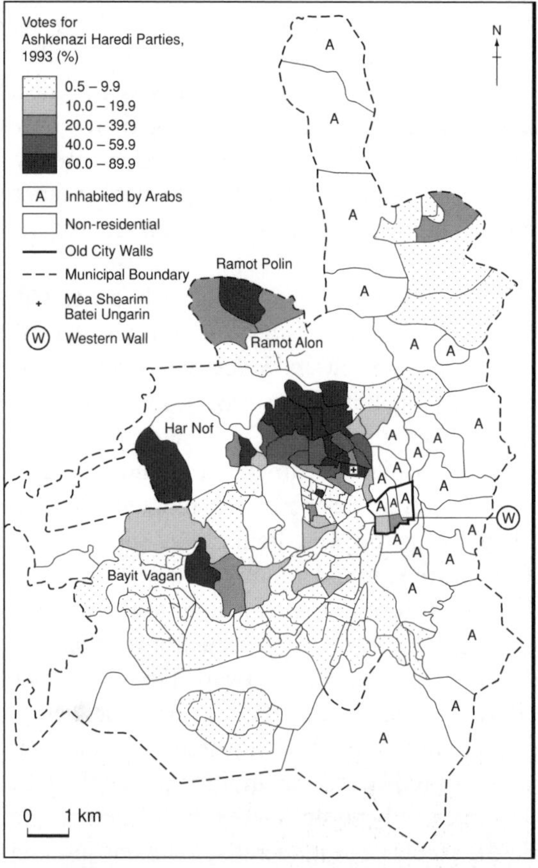

Figure 6.1 *Distribution of the Haredi (ultra-orthodox) population of Jerusalem in November 1993, as indicated by votes for Ashkenazi Haredi parties in local elections. (Source: Israel Central Bureau of Statistics; based on Map 1 in Shilhav 1998: 4)*

extended as far as the modern suburbs of Ramot Polin in the north and Har Nof in the west, with an outlier of Bayit Vagan in the south-west. Mea Shearim, founded in 1874 as one of the first settlements outside the Old City walls, was built strictly according to the rules of the Torah, with its own synagogue, market and yeshivas for religious study (Armstrong 1996: 356). This is where Isaac Bashevis Singer's character Hertz Grein found that there were 'still Jews':

> They wear garments that bear witness from afar that they are God's servants. When one puts on a capote [black caftan], a fringed ritual undergarment, a prayer girdle, when one grows one's beard and sidelocks, when one studies the Gemara [Torah] (yes, really and truly the Gemara), one doesn't read secular books, or go to profane theatres, or have assignations with women. (I. B. Singer 1998: 546)

To a writer on Jerusalem:

> Mea Shearim is a living museum of life-styles that existed for centuries among ultra-Orthodox Jews in the ghettos and shtetls of eastern Europe. Its energies are turned almost exclusively upon itself. Dressed still in the somber, confining clothes of an eighteenth-century northern-European ghetto, the residents of Mea Shearim inhabit [an] austere, long-vanished world . . . Nowhere can the ultra-Orthodox live out their customs more fully and more securely than here. It is a medieval world of poverty and unbroken faith – and of fanatical intolerance of other worlds of thought or ways of life. (Elon 1996: 185)

Exclusive identity is expressed most distinctively in dress, but perhaps most forcefully in posters used by rabbis to warn people of various dangers of secular society (Figures 6.2 and 6.3). More generally:

Figure 6.2 *The old ultra-orthodox Jewish quarter of Mea Shearim in Jerusalem. (Source: Author's collection)*

Figure 6.3 *Moral markers: posters prescribing local codes of conduct in Mea Shearim. One concerns the Sabbath and its desecration in public spaces; another is about the evils of allowing the Internet into the house, thus putting temptation in the way of children who are otherwise shielded from these sorts of images in their own segregated spaces (interpretation courtesy of David Newman). (Source: Author's collection)*

> The haredi areas in the northern part of the city have a character of their own, marked by social insularity, homogeneity with respect to the residents' commitment to religion, a high concentration of religious institutions and services intended solely for the haredi population . . . the haredi area delimits the boundaries of culture, and hence also permissible and prohibited behavioral patterns. Within these boundaries, certain kinds of behavior are legitimized and others are delegitimized.
>
> (Hasson 1996: 10–11)

Four cultural groups may be identified in the Jewish population of Israel, differentiated according to their approach to religion: secular, traditional, orthodox and ultra-orthodox or Haredi (Hasson and Gonen 1997). Like the secular, the traditional and orthodox tend towards moderation and the acceptance of other views on common life in the city. While modern Jewish orthodoxy is characterised by a commitment to the principles of *halacha* (Jewish law), Haredi ultra-orthodoxy favours the strict observance of tradition and custom that developed in the Jewish communities in Eastern Europe, with a tendency towards the stricter options of *halachic* rule (Shilhav 1998: 1). This requires not only special institutions but also distinctive forms of dress, strict observation of the

Sabbath including the closure of streets and prohibitions on trading, and, above all, religious conformity. There are clearly prescribed gender roles, with privileged practice of religious study for men while women are assigned a supportive role, in both cases with no choice but the difficult route of exit from communities the boundaries of which are diligently policed. Women and girls are held responsible for male sexual self-control, which requires 'modest' forms of personal presentation. This way of life is vigorously defended by the communities concerned, and is heavily subsidised by the state. The secular city at large is regarded as hostile by the Haredim; the influences of the modern are 'threatening the sacred space that the haredi community is trying to fashion for itself' (Hasson 1996: 10), protected by a 'wall of holiness' (Shilhav 1998: 120).

The Haredim seek to combat the threat to their way of life by various means. As they expand their residential space into new neighbourhoods, they attempt to control schools and other institutions, and even local government. They try to extend their prohibitions on the use of public space beyond their own neighbourhoods, for example by campaigns for the closure of other streets, most notably the major thoroughfare of Bar-Ilan Street which connects northern parts of the city with the main exit to the west. They oppose relaxations of constraints on Sabbath trading. They assembled a quarter of a million demonstrators in February 1999 to protest Supreme Court rulings limiting ultra-orthodox influence in matters of civil law, including marriage and burials. And, within their own territory, dress codes may be enforced by self-styled 'modesty patrols' (as featured in Naomi Ragen's play *Minyan Nashim – A Quorum of Women*). Those seeking exit from the community may be restrained, and even beaten. The imposition of community norms in private space is thought to be reflected in an increase in violence against women and complaints by women in orthodox homes, reported in 1998. (Most of this paragraph is based on reports in the *Jerusalem Post* and *Israel Wire*).

A study of the ultra-orthodox in urban government shows how the power they are able to gain from spatial concentration has changed: 'from being a defensive community, they have emerged as an offensive group, striving to spread their message and impose their values wherever possible' (Shilhav 1998: 6). The Haredi community adopts 'a dialectic attitude to modernity': rejection of its culture and values but acceptance of its instrumental component (Shilhav 1993: 277). Attempts to extend their sphere of influence may be resisted by the existing population, for example in Ramot Alon where secular residents prevented attempts to close the road leading to their neighbourhood, built a swimming pool against Haredi objections, and struggled to maintain the secular character of the schools and local council (Shilhav 1996: 80). Such 'territorial opposition' is part of the spatial dimension of 'normative conflict' (Shilhav 1998: 65). The danger, as perceived by non-orthodox Jews, is that the potential numerical superiority of the Haredim may enable them to take political control of the city of Jerusalem at large, with the power then to impose aspects of their own way of life, a prospect which may be accelerated by the outward migration of secular Jews fearing such an outcome (Hasson and Gonnen 1997).

How might the ultra-orthodox community be judged, in the context of group rights? Responding to a claim that there is an obligation in a liberal society to support even those cultures that flout the rights of their individual members (Margalit and Halbertal 1994), Okin (1998: 672) argues that ultra-orthodoxy is more likely than an open and liberal culture to harm the interests of both male and female children, and concludes that 'its public support is unacceptable for both liberal and specifically feminist reasons'. Her grounds are the limitations of the predominantly religious education system, which restricts the choices and well-being of boys, and of the merely supportive role open to girls. If this judgement is accepted, it still leaves open the possibility of a strategy of non-intervention. But it is clear that the Haredi community could not survive without state subsidy, which provides more than half its income. The conclusion that appears to follow is that there should be no place for the Heradim in a liberal society.

But, is Israel itself a liberal society? Evidently not, according to one observer:

> [T]he Israeli status quo diverges substantially from the common denominator of all liberal democracies in its infringement of freedom of conscience (by permitting only religious marriage and divorce) [and] in its infringement of the right to marry by denying the option of interfaith marriage . . . The problematic status of women in religious law; and a series of 'religious' laws in which the state dictates to citizens a way of life in matters that all Western liberal democracies leave to the individual also constitute an infringement of freedom of conscience. (B. Neuberger 1997: 40)

Some of this is a direct reflection of the political power of the ultra-orthodox, as part of 'the politics of theological coercion' (Newman 1999). Within the particular context of Israeli society, the Haredim perform an important symbolic role, reaffirming a distinctive Jewish identity through their commitment to strict interpretations of religious law. This, in its turn, plays a crucial part in legitimising the State of Israel, and the privileged status of Jews within it. From this comes the secular gain and hence the difficulty of judging Haredi intolerance as beyond the pale.

However, there is a further dimension to this case. Not content with their own privileged status, the Haredim wish to impose their way of life on others, through the territorial strategy of cultural domination and political hegemony over the entire city of Jerusalem, mentioned above.

They also exercise significant influence in some other cities, in particular Tel Aviv, and on the national political scene where they can hold a balance of power through proportional representation in the Knesset (parliament). Shlomo Hasson comments on the conflict in Jerusalem as follows:

> [E]ach of the two communities, secular (liberal) and haredi, claims to have justice on its side and holds its rights to be both legitimate and incontestable. In each case, however, the notions of justice and right are defined in the context of contrary sociopolitical cultures: the liberal-democratic culture of Western secular society and the haredi culture governed by Jewish law. Hence, there is no *a priori* criterion to adjudicate the conflicting claims of the two cultures and to determine whose views and rules should govern the functioning of the city. In the absence of such an *a priori* criterion, the power each community wields and its ability to mobilize and fight for its rights determines how Jerusalem functions. (Hasson 1996: 17)

However:

> In general the secular population will assert that what the haredi community is allowed in its districts cannot be denied the secular population. It will argue that to do otherwise would be unjust and contrary to the ethical principle of not doing unto others (the secular population) what you would not have done to yourself (undermining collective group rights). The secular population will claim that, following the haredi precedent, it, too, has the right to live its lifestyle without being subject to conflict, coercion, or duress. (Hasson 1996: 50)

The crucial and deciding point seems to be that Haredi intolerance cannot be universalised: they cannot claim their right to their own life in their own territory and at the same time deny this right to others, within a consistent moral discourse. Their claim to hegemony ultimately rests on its religious basis trumping other rights, for which there is no ground except faith. If there is a place for the Haredi, it is within a Haredi fundamentalist state. Or, perhaps, it is in competition with other fundamentalisms, in which context the following observations may provide a pointer:

> In Islam and Judaism, the religion is based not only, and perhaps not even primarily, on a personal, emotional, or mystical experience of the individual believer or on the tenets of his faith, but on a normative-systematic arrangement of the entire course of life, to the point that the religion is unwilling to recognize the separate existence of ethics and law. (A. Levontin, quoted by B. Neuberger 1997: 25)

It is an irony of geography and history that these two unyielding versions of the good may find themselves in competition for space, with no higher morality than might is right to offer a resolution.

CHAPTER 7

DISTRIBUTION: TERRITORIAL SOCIAL JUSTICE

> Although it celebrates sameness and cuts across differences, the postulate of moral equality is more elevating than humbling. (Selznick 1992: 491)

Like minority rights, social justice provides a powerful discourse for the protection and advancement of the vulnerable. Social justice is conventionally considered to focus on distribution, on how the benefits and burdens involved in social life are distributed among persons or groups, on the process involved, and on criteria which may be applied to its moral evaluation. The term 'territorial social justice' is sometimes adopted when the issue is explicitly that of distribution in geographical space, among territorially defined populations. After introducing some implications of difference for social justice, this chapter recognises the significance of the place of good fortune in people's lives, and then proceeds from recognition of human sameness, through needs and rights, to a general conception of (territorial) social justice. It concludes with observations on the universality of the concept of social justice, and on its particularity in practice. (This chapter draws on D. M. Smith 2000a, 2000b.)

DISTRIBUTION AND DIFFERENCE

Geographical engagement with social justice goes back three decades. There was a defining moment at the annual meeting of the Association of American Geographers in 1971, when David Harvey read a paper entitled 'Social justice and spatial systems' (first published as Harvey 1972). He identified these conditions which would have to be fulfilled for 'a just distribution justly arrived at' (Harvey 1973: 116–17):

> (1) The distribution of income should be such that (a) the needs of the population within each territory are met, (b) resources are so allocated to maximise interterritorial multiplier effects, and (c) extra resources are allocated to help overcome special difficulties stemming from the physical and social environment.

(2) The mechanisms (institutional, organizational, political and economic) should be such that the prospects of the least advantaged territory are as great as they possibly can be.

Income was broadly conceived as some measure of command over society's scarce resources. Prioritising the prospects of the least advantaged reflected the theory of justice recently set out by John Rawls (1971), to which we will return.

Harvey recognised that a more detailed examination of these principles would be required to build spatial theory around them. However, the subsequent development of human geography revealed little progress in this direction, despite a massive accumulation of literature on social justice in other fields (see Kymlicka 1990; D. M. Smith 1994a). So, when Harvey (1996) returned to the subject at book length twenty-five years on, his central question of 'the just production of just geographical differences' was virtually the same as before.

However, the language of the times had changed, to reflect the contemporary preoccupation with difference. Identifying those differences among persons (and groups) which are morally significant to the distribution of benefits and burdens has always been a central issue in social justice. In recent years a politics of difference has challenged ways in which people may be treated unfairly on the grounds of disability, ethnicity, gender, race, sexual orientation and the like, underlining the significance of social relations like domination and oppression and generating a critique of preoccupation with distribution (I. M. Young 1990b). This movement involved calls for recognition on the part of those whose experiences were excluded from the supposed universal 'we' based on the particular world-view of the Western male, bourgeois subject (McDowell 1995: 285). The claim for recognition itself may be considered a matter of justice, beyond the restrictive scope of the distributional paradigm, but often also a means to the more material end of economic equality.

The dual significance of difference, as a source of inequality and of solidarity against injustice, has been especially appreciated within feminism. Tension is also acknowledged between the risk of essentialism implicit in such categories as 'woman', mobilised in opposition to gender inequality, and the acceptance of different experiences of being a woman (or member of some such group of 'others'). Problems arising from recognition of increasingly complex, multiple or hybrid identities have led some feminists to advocate strategies for avoiding 'the postmodern nightmare' of a world in which there is nothing but diversity (Kobayashi 1997: 6). Acceptance of some common features of particular groups, and possibly of all persons, provides grounds for a restricted or strategic form of essentialism as a basis for political action. To thus place limits on the significance of difference is to define who belongs: 'who stands within, and without, the moral shelter of the same' (Kobayashi 1997: 4). Resolution of the tension between difference and sameness requires recognition that the rationale of the politics of difference is so that the different become part of a 'we' which is a source of social unity (Sypnowich 1993b: 106–7), as a community or nation but potentially including all of humankind.

The focus on difference has thus broadened the scope of social justice, and drawn attention to the disadvantage of specific groups. But in the process there has been an erosion of the sense of human sameness, or close similarity, underpinning the ideal of moral equality crucial to a just society. While recognition of salient forms of difference has helped to counter oppressive aspects of a universalising modernism, some of the greatest struggles for social justice in recent history (e.g. for black civil rights in the USA and against apartheid in South Africa) were more a case of the universalist notion of equal moral worth countering particular social constructions of difference. This highlights the way in which a geographical perspective on social justice has to work between universality and particularity.

Another important development since Harvey first wrote has been the consolidation of theoretical diversity. He recognises a plurality of theories – libertarianism, utilitarianism, contractarianism, egalitarianism and so on – which are 'equally plausible', finding the question of 'the most socially just theory of social justice' as intractable as ever (Harvey 1996: 398–9). Contrary to the postmodern suspicion of overarching frameworks, he argues that the powers or particularities that might favour one theory or another have therefore to be transcended, to find a discourse of universality and generality uniting social and environmental justice.

An important (re)discovery of the past quarter of a century of moral philosophising has been the interdependence of social justice and the good life (D. M. Smith 1997a). If we accept that the distinction 'is not given by some moral dictionary, but evolves as a result of historic and cultural struggles' (Benhabib 1992: 75), then to the debate over alternative theories of social justice must be added the deeper question of what might comprise a good way of living for humankind. This raises a further aspect of the significance of difference, for different cultures or ways of life value things in different ways (as demonstrated in the previous chapter), calling into question the possibility of a universal theory of how they should be distributed. Thus, 'we are distributing lives of a certain sort, and what counts as justice in distribution depends on what that "sort" is' (Walzer 1994: 24).

The emergence of the politics of difference, the persistence of theoretical diversity and recognition of the plurality of concepts of the good are crucial features of the intellectual context within which deliberations about social justice now take place. To these should be added a reluctance to focus heavily on the territorial dimension, in the light of the evident interdependence of the spatial and the social reflected in the rejection of 'spatial fetishism' in human geography as a whole. However, none of this rules out attempts to construct a general perspective on social justice with due attention to difference, with which to engage the changing world. Indeed, such a strategy seems more important than ever, if the power of the discourse of social justice is not to be blunted by undue reverence for particularity. And none of this seriously invalidates the unease about inequality which has traditionally motivated concern for social injustice, but which is being side-lined in the politics of these postsocialist times. The argument which follows lays some foundations for the perspective of social justice as equalisation (introduced in D. M. Smith 1994a).

THE PLACE OF GOOD FORTUNE

Understanding an egalitarian perspective on social justice begins with recognition of the place of good fortune (following D. M. Smith 2000a). This expression incorporates three meanings of 'place': the role or part played by good fortune in people's lives; position in some social structure, and place (or territory) in its geographical sense. Each has a bearing on human well-being, with important moral implications.

Interest in this issue can be traced back to the ancient Greeks. As Bernard Williams (1985: 5) puts it:

> Impressed by the power of fortune to wreck what looked like the best-shaped lives, some of them, Socrates one of the first, sought a rational design of life which would reduce the power of fortune and would be to the greatest possible extent luck-free.

In those hazardous times, it was recognised that achievement of the good life might not be entirely a matter of individual volition. He points out that most personal advantages and admired characteristics are distributed in ways which cannot be regarded as just, and that some people are simply luckier than others; morality is a value that transcends luck, and which has played a part in mobilising power and social opportunity to compensate for misfortune (Williams 1985: 195).

The role of luck re-emerged in recent times in arguments about desert, central to the liberal egalitarian perspective on social justice elaborated by John Rawls (1971). He began with the conventional system of 'natural liberties' in which careers are open to the talented, with all persons having equal opportunities in the sense of legal rights. However, there is no attempt to promote equality in background social conditions; far from it:

> [T]he initial distribution of assets for any period of time is strongly influenced by natural and social contingencies. The existing distribution of income and wealth, say, is the cumulative effect of prior distributions of natural assets – that is, natural talents and abilities – as these have been developed or left unrealized, and their use favored or disfavored over time by social circumstances and such chance contingencies as accident and good fortune. (Rawls 1971: 72)

The injustice of such a system is that it permits access to positions of advantage and distributive shares to be influenced by factors which are arbitrary from a moral point of view.

He therefore invokes the principle of 'fair equality of opportunity', under which persons with the same talent and ability and the same willingness to use them should have the same prospects regardless of their initial place in the social system. However, this conception also appears defective:

> [E]ven if it works to perfection in eliminating the influence of social contingencies, it still permits the distribution of wealth and income to be determined by the natural distribution of abilities and talents . . . distributive shares are decided by the outcome of the natural lottery; and this outcome is arbitrary from a moral perspective. There is no more reason to permit the distribution of income and wealth to be settled by the distribution of natural assets than by historical and social fortune. (Rawls 1971: 73–4)

Erasing the distinction between what may broadly be regarded as environmental effects and natural attributes achieves 'democratic equality', which strongly suggests equality of outcomes. In effect, Rawls made all sources of differential occupational achievement morally arbitrary. There is no case at the most basic level of justification for anything except equality in the distribution of his primary goods of liberty and opportunity, income and wealth, and the bases of self-respect. Thus: 'No one can be said to deserve anything (in the strong, pre-institutional sense), because no one can be said to possess anything (in the strong, constitutive sense)' (Sandel 1982: 92–3).

This argument from arbitrariness features prominently in subsequent work on social justice. For example, Richard Miller (1992: 228, 240–1) points to the social determination of the effect of a difference in raw talent, to inheritance transmitting unequal competitive resources along family lines, and to inequalities guaranteed by the organisation of education and production. He concludes: 'disadvantages in resources for social advancement are associated with generally inferior economic situations. It is as if the gamblers with the least funds were also dealt the fewest cards'

(Miller 1992: 255). He also argues that the unchosen risks of market competition stand in need of justification (Miller 1992: 274).

Place in its geographical sense is readily added to the argument from arbitrariness:

> So much of what people achieve is a matter of being in the right place at the right time, of having good luck in family, teachers, friends, and circumstances, that no one is in a strong position to take *much* credit for the way their lives turn out. There's no such thing as a literally self-made man [*sic*]. And so any judgement of desert will have to look closely at where responsibility really lies. (Baker 1987: 60)

If the distribution of resources across the world is entirely fortuitous, 'it is morally unacceptable that people's lot in life should be determined by this accidental feature' (Jones 1994: 167). Brian Barry (1989: 239) postulates Crusoe and Friday on two different islands, working equally hard and skilfully but with differences in production due to one island being fertile and the other barren, asserting that 'if anything can be called morally arbitrary – not reflecting any credit or discredit on the people concerned – it is this difference in the bounty of nature'.

The distribution of resources includes those created by humankind, like the local infrastructure, as well as those of the natural environment. It does not take a geographer to recognise the inequity of unequal access to facilities, such as good schools (e.g. Barry 1989: 220, 221), and of fiscal disparities between local governments (e.g. Le Grand 1991: 108, 128). This is all part of the undeserved inheritance. Thus:

> No one earns the right to be born to a family living in a spacious house in Armonk, New York, rather than a family living on a straw mat in the slums of Calcutta. Yet the enormous differences at these starts include enormous differences in life-prospects, given the same innate capacities and the same willingness to try.
> (Miller 1992: 298)

The chance of birth in a particular place on the highly uneven surface of resources carries no greater moral credit than being born to a rich or poor family, male or female, black or white. And such initial advantage as arises from the place of good fortune is readily transferred to future generations, similarly devoid of moral justification. Furthermore, for most people the capacity to change their place, from poorly to richly endowed, may be as limited as it is to change their gender or skin pigmentation. Yet, insofar as right of access to unevenly distributed resources is constrained by the boundaries of nation states for example, as accidents of history, this source of inequality might be considered morally irrelevant (Jones 1994: 160). Despite the rhetoric of globalisation, most of the world's people live in closed worlds, 'trapped by the lottery of their birth' in their nation state as a 'community of fate' (Hirst and Thompson 1995). And within nation states, the quality of such services as health care and education may be so variable as to depend on what in Britain is referred to as a 'post-code lottery'.

The argument from arbitrariness, as outlined here, has attracted vigorous opposition. There is a reluctance on the part of critics to concede to natural, social or chance circumstances everything about the individual, including responsibility for chosen life plans. This worries liberals dedicated to the notion of individual autonomy, and is anathema to communitarians with thicker conceptions of human identity. If the effort they expend, like all their other capacities, is only the arbitrary gift of nature or nurture, 'while its purpose is to leave us with persons of equal entitlement, it is hard to see that it leaves us with *persons* at all' (Walzer 1983: 261).

Those unwilling to assign everything to morally arbitrary fortune face the challenge of how to draw the line among attributes. There are enormous conceptual and technical difficulties in sustaining particular cuts between circumstances for which persons cannot be held responsible and those for which they can (Roemer 1996). The notion of justice as equality of fortune is also subject to the criticism that it is unduly concerned with individual circumstances and preferences, allowance for which can generate uncomfortable outcomes – like compensating persons for expensive tastes as well as for disabilities (E. S. Anderson 1999).

Robert Nozick (1974) opens up another line of critique. He argues that persons have the moral right to use such natural endowments as intelligence and skill to their advantage, providing that this does no harm to others: the thesis of self-ownership. Similarly, persons are entitled to hold and benefit from natural resources, provided that they got them justly by initial acquisition or by transfer (i.e. gift, inheritance or purchase). His criterion for the justice of initial acquisition is that no other persons are thereby made worse off (Nozick 1974: 178), a modification of the proviso of John Locke that an individual is entitled to appropriate natural resources providing that there is as much and as good left in common for others. Objections to Nozick's entitlement theory include the difficulty of demonstrating that no-one is worse off as a result of particular private ownerships of natural resources, and of tracing acquisition back through a series of transfers which, if unjust (e.g. involving deception, robbery or coercion), should be rectified.

An important issue arising from Nozick's similar treatment of natural endowments and acquired holdings is whether they may be different in some significant sense. Certainly, the ownership of the external world can deprive others, while ownership of one's body cannot (Reiman 1990: 173–5). And people cannot change their entire body, but may be able to change their location. As to the supposition that groups of people are entitled to monopolise the resources of the territory which they happen to occupy, encouraged by the concepts of national citizenship and sovereignty (see Chapter 6):

> [F]or the community as a whole to deserve the natural assets in its province and the benefits that flow from them, it is necessary to assume that society has some pre-institutional status that individuals lack, for only in this way could the community be said to possess its assets in the strong, constitutive sense of possession necessary to a desert base. (Sandel 1982: 101)

And without this, to rephrase Sandel (1982: 92–3) in the individual context, no group anywhere deserves anything. A community, or nation, might claim a right to land in which they have mixed their labour, and even their blood, and to the advantages to be derived therefrom. A similar argument might be applied to the physical infrastructure, built to give future generations as well as present people a better life. But this would still leave unanswered the possible injustice of initial acquisition and the moral arbitrariness of the good fortune of inheriting favourable conditions for sustaining a good life.

HUMAN SAMENESS, NEEDS AND RIGHTS

> Egalitarianism ought to reflect a generous, humane, cosmopolitan vision of a society that recognizes individuals as equals in all their diversity.
>
> (E. S. Anderson 1999: 308)

It follows from recognition of the place of good fortune that there is a strong case for equality, by persons and territorially defined population aggregates, or at least for narrowing the gaps which have arisen from morally arbitrary factors. The crucial question is: equality of what? Should it be opportunities, resources, capabilities, welfare outcomes or even happiness? This question is complicated by the fact that the individual freedom to choose life plans so revered by liberals means that everyone might require a unique bundle of goods. As already noted, difference appears to work against some common conception of the good and of what is required to attain it.

However, these problems may be largely circumvented by recognising that what people actually require for life is much the same, whoever and wherever they are, because they are themselves much the same. The common proposition that persons are 'born equal', or even 'equal in the sight of God', captures the powerful notion of moral equality, often associated with Immanuel Kant, and invoked in everyday practices of human reciprocity (see Chapter 2). Thus:

> The first premise of moral equality is that all people are of the same kind, by which we mean that they are alike in morally relevant ways ... The norm of respect for all humans as persons is based on special facts about human nature.
>
> (Selznick 1992: 483)

These common elements of human nature form the empirical foundations of morality as other-regarding conduct, based on mutual recognition of sameness: 'they tell us who is that "other" for whom we properly have concern' (Selznick 1992: 484).

Any suggestion these days that there may be such a thing as human nature attracts suspicion of essentialism. 'Any definition of a human nature is dangerous because it threatens to devalue or exclude some acceptable individual desires, cultural characteristics, or ways of life' (I. M. Young 1990b: 36). However, deciding what may be 'acceptable' requires standards capable of transcending the here and now of specific individual, group or local practices. 'Differences cannot fully flourish while men and women languish under forms of exploitation; and to combat those forms effectively implicates ideas of humanity which are necessarily universal' (Eagleton 1996: 121). As to the assertion that the common traits of human beings are not substantial enough to constitute a useful notion, their powers are, by comparison with other terrestrial species, 'truly formidable' (Geras 1995: 66).

Recognition of the significance of human similarity can be found in contemporary geography. For example:

> [H]uman beings have much in common. We live in a concrete material environment and we share basic biological, social, intellectual, and perhaps even spiritual capacities; we also share the capacity to reason. Losing sight of this basic reality comes from too great an emphasis on difference and diversity. (Sack 1997: 4)

While acknowledging the significance of difference, Harvey (1996: 130) emphasises the importance of human similarity in alliance formation between seemingly disparate groups, in re-establishing 'a conception of social justice as something to be fought for as a key value within an ethics of political solidarity built across different places'. Thus, there are moves to recover a sense of human sameness, but without abandoning insights gained from understanding the particularity of persons and places.

The next step is to consider human needs. The notion of need implies the moral strength of some authority external to the individual, as opposed to subjective personal want or desire. There are also implications of entitlement, so that what someone needs is not earned or deserved: 'common humanity is reason enough for a claim on another's superfluity' (Ignatieff 1984: 35). Some needs may be referred to as basic, to stress their urgency and thereby give them special moral force. The relativist claim that human needs are contextual, specific to particular times, places and cultures, may be countered by the universalist argument that all persons share the same basic needs.

However, attempts to define universal needs reveal differences. For example John Kekes (1994: 49) identifies what are claimed to be context-independent requirements for human welfare, set by universal, historically constant and culturally invariant needs created by human nature:

> Many of these needs are physiological: for food, shelter, rest, and so forth; other needs are psychological: for companionship, hope, the absence of horror and terror in one's life, and the like; yet other needs are social: for some order and predictability in one's society, for security, for some respect, and so on.

Compare this with the more restrictive view of Onora O'Neill (1991: 279):

> It is not controversial that human beings need adequate food, shelter and clothing appropriate to their climate, clean water and sanitation, and some parental and health care. When these basic needs are not met [human beings] become ill and often die prematurely. It is controversial whether human beings need companionship, education, politics and culture, or food for the spirit – for at least some long and not evidently stunted lives have been lived without these goods.

In short, what is required for a human life is subject to interpretation.

One response is to define universal human needs in a minimal sense, related to immediate physical survival. However, the essential requirements for living day-to-day, as with the avoidance of illness or premature death, would hardly differ from those of a non-human creature. If the notion is that of *human* needs, then there should be something distinctively human about them. The argument for a broader definition rests on propositions concerning the distinctive nature of being human rather than any sentient creature.

Len Doyal and Ian Gough (1991: 37) are closer to Kekes than to O'Neill in asserting that our mammalian constitution shapes needs for such things as the food and warmth required to survive and maintain health and that our cognitive attitudes and experience of childhood shape needs for supportive and close relationships. Their hostility to relativism is expressed in the notion that all people share one obvious need: to avoid serious harm. This goes beyond failure to survive in a physical sense to include impaired participation in the prevailing social milieu. From this follow two basic needs (in their terms): for the physical health to continue living and functioning effectively and for the personal autonomy or ability to make informed choices about what to do and how to do it in a given societal context. The actual need satisfiers, in the form of goods and services, may be culturally specific, as opposed to the universality of the basic needs themselves. This is similar to the approach adopted by Amartya Sen (1992) to poverty, as absolute or universal in the sense of impairing people's capability to function but relative with respect to the commodities required to alleviate it. That there is more to human life than mere physical survival or even longevity is endorsed by others who claim to derive sets of needs from human nature or the requirements for human flourishing (e.g. Brown 1986: 159; Griffin 1986: 86–7; Nussbaum 1992: 222).

In a critique of the equality of fortune perspective, Elizabeth Anderson (1999: 314) stresses the identification of 'certain goods to which all citizens must have effective access over the course of their whole lives', and without resort to paternalism. She follows Sen's capabilities approach in highlighting three aspects of human functioning: as a human being; as a participant in a system of cooperative production; and as a citizen of a democratic state. These require: access to means of sustaining biological existence and basic conditions of human agency; access to means of production, education, freedom of occupational choice; the right to make contracts and enter into agreements with others; the right to receive fair value for one's labour, and recognition by others of one's productive contribution; rights to political participation such as freedom of speech and the franchise, and also access to the goods and relationships of civil society (E. S. Anderson 1999: 317–19). Her emphasis on the importance of people being able to avoid or escape oppressive social relationships is a reminder of the limitations of the narrowly distributive approach.

The basic needs perspective, in theory and in development policy and practice (see Chapter 8), further strengthens arguments from essentialism by adding attention to context. For example, Martha Nussbaum (1992: 205) finds that 'legitimate criticism of essentialism still leaves room for essentialism of a certain kind: for a historically sensitive account of the most basic human needs and human functionings'. This is a case of the strategic form of essentialism advocated by some feminists. Nussbaum (1992: 212) suggests that to give up on all evaluation, and in particular on a normative account of human being and functioning, is to turn things over to a world situation in which the forces affecting the lives of women, minorities and the poor 'are rarely benign'. She stresses that, although the capabilities approach advances cross-cultural norms, this universalism 'derives support from a complex understanding of cultures as sites of resistance and internal critique' (Nussbaum 1998: 770).

The conclusion is as follows:

> We cannot jettison essentialism because we need to know among other things which needs are essential to humanity and which are not. Needs which are essential to our survival and well-being, such as being fed, keeping warm, enjoying the company of others and a degree of physical integrity, can then become political criteria: any social order which denies such needs can be challenged on the grounds that it is denying our humanity, which is usually a stronger argument against it than the case that it is flouting our contingent cultural conventions. (Eagleton 1996: 104)

Even the difference theorist recognises that justice 'requires a societal commitment to meeting the basic needs of all persons', for if they suffer material deprivation, 'they cannot pursue lives of satisfying work, social participation, and expression' (I. M. Young 1990b: 91).

All this suggests the universality of some general conception of what is needed to sustain a distinctively human form of life, along with a degree of cultural relativity in how it is interpreted and satisfied. But it would be surprising if the goods and services required differed very much, at the relevant level of living endured by the world's most needy. 'Relief workers in Africa don't have to probe deep philosophical questions to discover that certain things are needed: those needs are immediate and obvious' (Baker 1987: 15). As Nelson Mandela (1994: 293) found, travelling beyond South Africa, 'poor people everywhere are more alike than they are different'. An argument for equalisation of the same or a closely similar package of the means of basic need satisfaction follows from the observation of human sameness or close similarity. Very simply, 'everyone has similar needs that must be met before any sort of worthwhile life can be lived' (Belsey 1992: 47).

The moral argument for distribution according to need has a long history, going back at least to Karl Marx. If certain things are needed to live a human life, then it might be argued that all persons everywhere should have them by right. Again, human sameness may be invoked, for a distinctive feature of the rights perspective is 'that it is based on what human beings have in common, their common needs and capacities, and on a belief that what they have in common is more important that their differences' (Almond 1991: 266). If social justice is to prevail, then the moral imperative often associated with rights can give strength to specific entitlements. However, the notion of rights raises difficult issues, with respect to what they are, how they should be prioritised, who bears them (and where), and who have the consequent obligations to ensure that the rights are fulfilled (see for example Jones 1994).

A fundamental difficulty is that what may be labelled for simplicity as liberty rights and welfare rights can conflict. And the former may be easier to handle than the latter. With respect to liberty, every individual is a rights holder, while everyone is also obligated not to interfere with the liberty of others. The necessary institutional arrangements are based on laws for the specification and protection of the actual liberties to which people are entitled. However, although the bearers of welfare rights are all people as individuals, it is not clear who is to meet the claims implied. For example, if all people have a right to the food, clothing and shelter necessary to survive, or to other things so that they may flourish, who is obliged to provide, and at what geographical scale? Is it other particular individuals or groups (e.g. the family or community), or local government, or national state, or some international agency? It is these questions which make welfare rights so much more difficult than liberty rights to define, and to claim.

One ingenious resolution has been offered by James Sterba (1981). He conceives the right to life as both a positive right (i.e. to the satisfaction of a person's basic needs required in order not to seriously endanger health or sanity) and as a negative right which requires that everyone in a position to do so does not interfere with a person's attempts to meet basic needs. This raises the question of whether persons with goods and resources surplus to their basic needs are justified in prohibiting others less well endowed from using them for their basic needs. He concludes: 'For most people their right to acquire the goods and resources necessary to satisfy their basic needs would have priority over any other person's property rights to surplus possessions' (Sterba 1981: 102). And in this view he is not alone, for even such a staunch defender of private property as John Locke believed that all have a right to physical subsistence which overrides the property rights of others (Dunn 1984: 43). This has important implications for the institutions necessary for effective redistributive social justice. However, universal welfare rights remain a fragile basis for social justice in practice, and patently ineffective for most of the world's population, without the moral force, political will and economic resources required for their fulfilment.

SOCIAL JUSTICE

> Justice . . . is not ever an end in itself, but a means to some further end, some conception of the human good which does not inhere merely in the distribution of scarce goods. (Greaves 1994: 34)

That some people in some places are better off than others elsewhere is an outcome of geography as well as of history. If the forces creating these patterns of inequality carried their own moral justification, there would be no problem of social or territorial (in)justice. However, so powerful are the arguments for equality that 'the central issue in any theory of justice is the defensibility of unequal relations between people' (Barry 1989: 3). Rather than being based on conflicting ultimate values, all contemporary theories of the just society have the same foundation: equality (Kymlicka 1990: 4–5). Thus, 'the major ethical theories of social arrangements all share an endorsement of equality in terms of *some* focal variable, even though the variables that are selected are frequently very different from one theory and another' (Sen 1992: 3). This difference arises in part from competing conceptions of the good, which are hard to separate from what is right or just.

It requires strong moral argument to defend inequality in access to a good life, by territory or any other criterion. While the desert associated with making great efforts in the form of contributions to society, perhaps overcoming particular obstacles, could carry special weight (as in Harvey's original formulation), even this capacity might be traced to the chance of genetic endowment, environment or socialisation. The defence of inequality adopted by Harvey (1973) reflected what John Rawls referred to as the difference principle. This requires social and economic inequalities to be arranged so that they are 'to the greatest benefit of the least advantaged', all social primary goods to be distributed equally unless an unequal distribution is 'to the advantage of the least favoured' (Rawls 1971: 302–3). Even if the place of good fortune is taken to undermine the moral credit for most if not all individual achievement, the difference principle is a defensible concession to the possibility that inequality can work to the advantage of everyone and especially the poor. The most obvious case is that of rewarding the more efficient (individuals or territories), in order to increase aggregate production from which the poor gain.

The continuing appeal of Rawls's difference principle is reflected by others. For Doyal and Gough (1991: 132): 'inequalities will be tolerated to the extent that they benefit the least well off through leading to the provision of those goods and services necessary for the optimization of basic need-satisfaction'. The disparities that arise from barriers to advancement and such benefits as inheritance under capitalism lead Miller (1992: 187) to require that inequality works in the interests of the worst-off, his moral defence being 'equality of opportunity, which every partisan of social freedom would accept on adequate reflection'. And the conception of alternative development advocated by Friedmann (1992), stressing the moral claims of the disempowered poor (see next chapter), has a distinctly Rawlsian tone. The best-known derivation of the difference principle is that this is what persons would agree to from behind a 'veil of ignorance' as to their position in society, so that if they were to end up among the worst-off they would be as well-off as possible. However, it is more plausibly defended by the argument from arbitrariness explained above (Barry 1989: 225–6). It can be posited against the difference principle that, once all have their basic needs met, inequalities that fail to benefit the worst-off may not be morally objectionable. But this would depend on how such inequalities feed back into society, influencing prevailing conceptions of the good, including social justice. Given the ease with which wealth buys economic, political and cultural power, the worst-off need the best protection moral theory can devise.

It remains to consider the priority which should be given to equalisation of sources of human need satisfaction. As was recognised above, the right to material well-being might conflict with other moral values, like liberty. Rawls's first principle is that each person is to have 'an equal right to the most extensive total system of equal basic liberties compatible with a similar system of liberties for all', and this takes precedence over his second principle that inequalities are to be arranged to the greatest benefit of the least advantaged and attached to offices and positions open to all under fair equality of opportunity (Rawls 1971: 302). This order of priorities reflects liberalism's implicit theory of the good.

However, liberty does not have to be given priority over social and economic equality. A reformulation from a Marxian perspective proposes the following first priority:

> Everyone's basic security and subsistence rights are to be met: that is, everyone's physical integrity is to be respected and everyone is to be guaranteed a minimum level of material well-being including basic needs, i.e., those needs that must be met in order to remain a normal functioning human being. (Peffer 1990: 14)

This takes precedence over a maximum system of equal basic liberties, equal opportunity, and an equal right to participate in social decision-making. A version of the difference principle is specified as: 'Social and economic inequalities are to be justified if and only if they benefit the least advantaged . . . but are not to exceed levels that will seriously undermine the equal worth of liberty or the good of self-respect' (Peffer 1990: 14). The priority given to economic and social security over liberty allows such hallowed tenets of liberalism as private property and freedom from imposed conceptions of the good to yield to the basic needs of the worst-off. This could be a morally superior response to the reality of the contemporary world, in which most people have some elementary liberties guaranteed under the law (though not necessarily upheld in practice), while this is not true of rights to even the most basic means of subsistence. The question of what kind of liberty some people in some places actually enjoy, if their major preoccupation in life is to survive rather than to flourish, might add weight to the prioritisation of satisfaction of material needs at some expense to individual liberty.

Of course, to equalise the wherewithal for subsistence, or basic need satisfaction, on a global scale might leave little surplus to sustain local indulgences. In other words, given limits to global resources, satisfying everyone's basic needs here and now, never mind provision for future generations, greatly limits the scope for inequality (Sterba 1981; 1998: 63). Such a strategy would have implications for the good life, which would exclude those excesses of luxury consumption currently enjoyed by a small minority of the world's population at the expense of the more modest needs of the vast majority. The wider the spatial scope of (re)distribution, as well as the more generous its conception of need, the more severely egalitarian its consequences. And the more egalitarian the outcomes, the greater the limitations on individual or group conceptions of the good which require disproportionate shares of sources of need satisfaction. This underlines the interdependence of justice and the good life.

UNIVERSALITY AND PARTICULARITY

> [A] theory of justice cannot simply be a theory of what justice demands in this particular society but must be a theory about what justice is in any society. (Barry 1995: 6)

> Abstract principles can guide context-sensitive judgement without lapsing into relativism. (O'Neill 1992: 53)

A conception of social justice has been constructed above, beginning with the moral arbitrariness of sources of inequality (especially those associated with place or territory), and then moving from recognition of human sameness to basic needs and to their satisfaction as of right. From the initial identification of inequality in the necessities for human life, social justice requires a process of equalisation over time, as expeditiously as possible, until the constraint of Rawls's difference principle is reached. Insofar as geographical space is one (important) dimension within which (re)distribution takes place, and is evaluated, the outlines of territorial social justice are revealed.

Further theoretical refinements, along with operational considerations, are elaborated elsewhere (D. M. Smith 1994a). The purpose of this concluding section is to return to the question of universality versus particularity in social justice, as an instance of this fundamental tension in the more general engagement of geography with morality and ethics. If the constrained egalitarian perspective sketched out here has claims to universality, what are they, and how are they justified?

The first point to recognise, in the light of the historical geography of morality explained in Chapter 2, is the contextuality of the very notion of justice and of its social variant. Agnes Heller (1987) explains how the ethical and political aspects of the traditional concept of justice were separated during the eighteenth and nineteenth centuries. The former came to constitute the modern field of ethics or moral philosophy, while the latter became focused on institutional arrangements rather than with the best possible moral world. The question of the best social world became largely a matter of the just distribution. The triumph of the right over the good in the attempt by Rawls (1971) to replace the teleological theory of utilitarianism by a deontological conception of justice as fairness is thus 'the shabby remnant of the "sum total of virtues" that was once called "justice"' (Heller 1987: 93). Heller's claim that the crisis of modern consciousness requires a new ethico-political concept of justice is part of the contemporary reconnection with the good life (D. M. Smith 1997a: 20). Even the exponent of the politics of difference, Iris Marion Young, commends ancient thought, which 'regarded justice as the virtue of society as a whole, the well-orderedness of institutions that foster individual virtue and promote happiness and harmony among citizens', referring to 'a postmodern turn to an enlarged conception of justice, reminiscent of the scope of justice in Plato and Aristotle' (I. M. Young 1990b: 33, 36).

Times change, and with them conceptions of justice, if there are any:

> [T]he question of social justice has a history. The question could not be widely entertained until people generally began to see their social structure as intentionally alterable. Thus, it only appeared as a live social question, not just as a matter of utopian fantasizing, around the time of the birth of capitalism and modern science, in the seventeenth century. By then, the printing press had already created a public consciousness, parliaments were actively tinkering with social relations, economic relations were coming unhinged from tradition and being remade according to freely undertaken contracts, and all of nature was a great machine to be made over in the service of human goals. (Reiman 1990: 226)

The possibility of rectification of injustice, via redistribution, was becoming conceivable as a moral project, thanks to developments within ethics, and feasible as a practical proposition, thanks to the changing scope of state power.

And with geographical changes, accompanied by changes in spatial awareness, came a growing sense of responsibility to distant others (traced in Chapters 5 and 6). As the problems of the world's poor 'peripheral' countries became more evident, and more evidently linked to their relationships with richer 'core' nations, they became part of the moral universe of economic development planning, with its offshoot of development ethics (see next chapter). Interdependence culminated in the globalisation of economic relations; international justice emerged as an issue (e.g. Attfield and Wilkins 1992; Thompson 1992), again because the morality became conceivable and redistribution at this scale feasible. As the discipline of geography was drawn into the discourse of social justice, the prefix 'territorial' came into play. The expanding spatial scope and seriousness of environmental pollution led to the infusion of an environmental component into justice and ethics (see Chapter 9). Thus the understanding of justice is subject to further and continuing change.

Universalism is so much part of our Enlightenment heritage that aspirations of this kind are understandable in theorising about social justice. How this stance might also accommodate particularity is explained by Onora O'Neill (1992; 1996). She distinguishes between idealised and relativised theories of justice, the former stressing the need to abstract from particularities, the latter conceding to them. 'The first move is to argue for abstract principles of universal scope'; justice in the first place is 'a matter of keeping to principles that can be adopted in any plurality of potentially interacting beings' (O'Neill 1992: 52, 64). This leads to the rejection of deception, coercion and other ways of victimising persons as universal principles of justice. Moving from abstract principles to determinate judgements leads her to ask 'to what extent the variable aspects of any arrangements that structure vulnerable lives are ones that *could have been refused or renegotiated by those they actually constrain*' (O'Neill 1992: 68). This bears a strong resemblance to a principle advanced by Brian Barry (1995: 7): 'it would widely be acknowledged as a sign of an unjust arrangement that those who do badly under it could reasonably reject it'. There is a distinctly Rawlsian ring to both formulations, with their assignment of the power of veto to the badly-off. But crucially, they deal in social relations, not merely distributions.

O'Neill (1996) goes on to show that the principle of injuring others cannot be universalised, because once injured a person cannot injure others, so to universalise the principle of non-injury follows. Thus: 'justice is in the first instance a matter of living lives and of seeking and supporting institutions and policies that reject injury', so 'commitment to universal principles of justice is most effectively expressed through specific institutions that limit risks of injury, so helping to secure and maintain basic capacities and capabilities for action for all' (O'Neill 1996: 179, 191). A similar argument leads to the conclusion that indifference to and neglect of others cannot be universalised, so some form of care and concern follows (as explained in Chapter 5).

After deriving these abstract universal principles, the second move takes account of 'the context and particularities of lives and societies' (O'Neill 1992: 53). O'Neill is prepared to recognise some significant differences in actual cases, but without conceding to idealisations which merely reflect accepted beliefs, traditions or practices. Her defence of abstraction is at the same time a rejection of the idealisation of agents implicit in the denial of salient differences, such as gender:

> Principles indeed abstract from differences: but it does not follow that they must assume idealized accounts of human agents that not merely bracket but deny human particularities and differences. Universal principles indeed apply across a plurality of differing cases, but need not prescribe or proscribe, recommend or reject rigid uniformity of action or entitlement. (O'Neill 1996: 77)

And her defence of universal principles against particularism is at the same time a recognition that they can be applied to particular situations without accepting relativism:

> [T]he diversity of situation and pluralism of belief found in different times and places do not undermine principles of justice. Principles of justice can be justified quite generally, and can be used to judge the specific constructions of justice which quite rightly, take different forms at different times and places.
>
> (O'Neill 1996: 173–4)

There is an alternative strategy to the one outlined by O'Neill, however. This is to begin with the recognition and observation of the full expression of difference, and to respect it – including different conceptions of social justice. However, this concedes to relativism: 'If all we can say to another person is, "Your conceptions of justice are true for you, in your cultural context, but mine are true in my context", meaningful debate about justice must cease' (Low and Gleeson 1998: 197). In such a situation, debates on justice as actually operated in South Africa under apartheid, for example, or in the ghetto under the Nazis, would be incapable of resolution. To undertake normative evaluation implies that there is some basis for comparison, and ultimately some kind of moral truth.

O'Neill's strategy may be compared with Michael Walzer's communitarianism, which might be expected to give greater weight to particularity:

> Justice requires the defence of difference – different goods distributed for different reasons among different groups of people – and it is this requirement that makes justice a thick or maximalist moral idea, reflecting the actual thickness of particular cultures and societies. (Walzer 1994: 33)

His reassertion of the significance of the local and particular, in the context of new or resurgent 'tribalisms', suggests that moral minimalism at the universal level leaves room for all the tribes, 'and so for all the particularist versions of justice' (Walzer 1994: 64). The danger is that this room may harbour practices which may be objectionable, for some of his thin or minimalist reasons: 'because we oppose oppression, deceit, and torture'. That these can be perpetrated by 'tribes', as well as under the totalitarian rule to which he takes exception, is a horrific fact of recent experience.

Building on communities is risky, as Milton Fisk (1995: 231) warns: 'there are all sorts of communities. Not all of them will provide a foundation for political morality in general or even justice in particular.' He sees the problem of justice as finding a way, beyond the contending schools of liberalism (universality) and communitarianism (particularity), of resolving conflicting claims so as to perpetuate the social bond:

> Justice, most would agree, builds bridges between individuals, between groups, between nations. It reaches out beyond any particular interest to draw in some potentially conflicting interest. In drawing these [two] particularities together, the just resolution acts like a universal. But paired with this universality is still a third particularity. For justice is done from a standpoint that does the drawing together. It is the standpoint taken up by some individual, some group, some nation. It doesn't escape into pure universality. The problem then arises as to how justice can be sufficiently universal so as to draw different tendencies together if it is never more than justice from a particular standpoint. (Fisk 1995: 222)

The perspective of social justice as equalisation, sketched out in this chapter, reflects some of the particularities of the world in which we now live, with its gross inequalities, residual egalitarian sentiments and abiding faith in grand projects for human betterment (actual experience and postmodern scepticism notwithstanding). Within this world, the perspective is general and flexible enough to do the initial work required of universal principles: to engage the specific manifestations of inequality in particular societies in a manner which facilitates comparison and generalisation. In the process, contextual thickening will take place, as understanding of the particular society is built up. The findings can then be related to others: 'past resolutions begin to lay a basis for formulating principles of justice that achieve a higher level of universality than what would be involved if cases were taken in isolation' (Fisk 1995: 227).

In the next chapter the perspective of social justice as equalisation will be deployed in a particular case, within the broader context of development, to reveal something of its scope and limitations.

CHAPTER 8

DEVELOPMENT: ETHICAL PERSPECTIVES

> The biggest challenge for development as good change in the long term, is to find more ways in which those with more wealth and power will not just accept having less, but will welcome it as a means to well-being, to a better quality of life.
>
> (Chambers 1997: 1751)

There are few greater challenges facing the contemporary world than those associated with development. A special issue of the South African journal *Theoria* celebrated the 'moral victory' marked by the end of apartheid, but offered this warning:

> [I]t is precisely at the moment of such victory that the multiple promises of modern developmenta projects – promises of equality, social harmony and genuine political freedom – will come to be seen in all their fragility and potential unredeemability. For the history of the twentieth century is in substantial measure a history not only of the success, but also of the dismal failure, of great experiments in modernization to deliver those goods – moral and material – encoded in the emancipatory visions by which they have been both guided and legitimated. (The Editors 1991: i)

The Editors go on to point out that the development process raises issues that are not only technical but also profoundly theoretical and moral. This severely taxes the various disciplines involved in development studies; among others, 'it urges geographers to pause and reflect upon its implications for the lived environment and the spatial organization of our public lives'.

The concept of development involves both a state and a process. The state of development of a society refers to the well-being of its population. The process of development refers to how this state was achieved, and how it is to be maintained and possibly improved. But such general definitions leave much to be clarified. How is well-being to be measured, with respect to what bundle of goods? Can this be compared among nations or regions, and if so by what (and whose) set of values? Or should the level of development of a society be left to culturally relative criteria, comparable only within the same society as it was at some point in the past (even if this can be done, given the possibility of criteria changing over time)? Should the state of development incorporate distribution of the means of well-being among individuals, groups and territories, as well as its aggregate level? As to the process of development, how is it most effectively achieved? Is there a universal model to be followed, or is the development process culturally and locally specific?

The normativity of these issues should be immediately apparent. So should their links to other issues encountered in earlier chapters, social justice being the most obvious. And it will not be surprising if some of the same underlying philosophical problems also impinge on the understanding of development: tension between the universal and particular, the cosmopolitan and communitarian, the global and local. This chapter introduces the moral content of development, in the context of what has become known as development ethics. Rather than elaborating an inevitably incomplete review of the disparate literature in this still emerging field (see Grimes 1999), the bulk of the chapter uses a case study of South Africa to explore some alternative development ethics.

INTRODUCING DEVELOPMENT ETHICS

It is only recently that philosophers have joined practitioners from other disciplines in giving serious attention to what development means. One points out that any definition of development is at the root evaluative, but identifies commonly recognised ingredients: change, which is the object of practical commitment, and which is of positive value or good, and which therefore ought to take place (Dower 1989). This raises first of all the questions of what kind of change is for the good, and which aspects of life it would be better not to change. Then there is the question of whether deliberation or intentionality is itself good, or even effective as means of inducing and guiding change. For example, there could be grounds for favouring spontaneity to the planning and control usually assumed to go with development. What may be good (or best) change is obviously related to the desired state of development, but any goal may be contested, in itself or in competition with others, growth versus equity being merely the most obvious conflict. And there can be value judgements involved in choice of means, given an agreed end, like preference for local community initiatives over those emanating from central government.

The promotion of development is thus a deeply normative project, connected with conceptions of human good. However, the identification of development problems and the practice of development planning tends to be largely a technical exercise, with the morality left implicit for the most part. The state of development is usually conceived very much in material terms, with development as a process still often prefixed by 'economic'. And insofar as culture may enter the scene, it is still all too frequently assumed that 'underdeveloped' nations or regions are impeded by their own ways of life, and that the sooner they 'modernise' the better. Modernise is usually a synonym for 'Westernise', which itself is shorthand for engaging in the competitive struggle of contemporary capitalism as the only possible prospect of surviving, never mind prospering, within the globalising marketplace.

It is against this background that an interest in development ethics has emerged in recent years. One exponent defines the field as 'the normative or ethical assessment of the ends and means of Third World and global development' (Crocker 1991: 457). He includes the following questions, among others: what should take place in Third World development, and what goals should poor countries pursue; how should development be defined, and should it be descriptive, prescriptive or both; how should the benefits and burdens of development be distributed; what responsibilities, if any, do rich countries, regions and classes have towards the impoverished; what international structures are called for by global justice; who ought to decide these kind of questions; what are the implications of political realism, moral scepticism and moral relativism for the possibility and practice of development ethics. A formidable agenda indeed.

A further, fundamental question of definition in development ethics involves the environment, implicit in such notions as sustainable development. A pioneer of the field calls for an approach that joins the two normative streams of a regard for 'environmental wisdom' and a concern for 'universal economic justice'; thus the task of development ethics is 'to formulate this ethic of integral development, locating environmental concerns at the heart of normative discourse on development' (Goulet 1990: 36; see also 1995). At the risk of failing to give this integration the respect it deserves, the approach adopted in this chapter is to focus on human aspects of development, reserving the environment for the next chapter. This best reflects an existing literature and professional practice which tends to distinguish between development ethics and environmental ethics.

The most convenient way of further introducing the subject matter of development ethics is through a brief review of changing development theory and practice. Such changes reflect the academic context within which development discourse is engaged, and the economic, social and political context within which development actually takes place. And in all this morality is deeply implicated, including the changing normative stance of the academy and the changing normative content of political economy.

The first really influential approach to development was that associated with stages of economic growth, popularised by Walter Rostow (1960).

This incorporated a notion of human progress as almost linear inevitability, given a particular kind of economic development, with the 'take-off' from traditional society to the age of 'high mass consumption' depicted almost with the reliability of jet propulsion. It was given spatial content by John Friedmann (1966), with considerable influence on the way in which geography engaged development issues (e.g. Fair 1982). Stages were recognised in the organisation or integration of the space economy, settlement pattern and network of transport and communication, the planning of which facilitated the diffusion of economic growth and the benefits of modernisation assumed to accompany it, as a 'trickle down' (or spread effect) from metropolitan core to rural periphery. In the process, regional disparities in living standards would be eroded, as at the international scale.

The basic assumptions of what soon became orthodox development theory were identified as follows (Foster-Carter 1976). Development did not involve irreconcilable interests, between developed and underdeveloped nations and between social groups; there are no structural connections between development in some places and underdevelopment in others; the modern is good, the traditional bad; development means becoming like the West. While those committed to the orthodoxy might claim that their perspective was empirically grounded, its normative (and ideological) content was clearly signalled by the sub-title to Rostow's book: 'A Non-communist Manifesto'. It proclaimed the near certainty of a particular conception of progress, based on the supposed natural process of economic growth under capitalism, as opposed to the spectre of central planning under socialism.

The perspective of development via economic growth was eventually challenged by what became known as dependency theory, elaborated in particular by Andre Gunder Frank (1971). He countered the basic assumptions of the prevailing orthodoxy, proposing that irreconcilable interests generate winners and losers in development, that the 'development of underdevelopment' is promoted by structural links between countries of the centre and those of the periphery, that what is modern is not necessarily good and traditional bad, and that real development does not (or should not) mean Western-style modernisation (Crocker 1991: 464). Thus, the underdevelopment of some parts of the world could be explained by their dependence on others. The orthodoxy's spatialised version, the so-called diffusionist paradigm, was also subject to critique (e.g. Browett 1980); its promised process of equalisation was shown to be empirically questionable and theoretically flawed, and its values biased in favour of the benefits of modernisation and against its costs.

Equity or redistribution with (or through) growth became the catch-phrase of an alternative development paradigm, expressing a new moral priority. The 1970s saw this consolidate as an emphasis on the satisfaction of basic needs. The World Bank and the International Labour Office became committed to this strategy, the latter producing a definition of basic needs to include the following: minimum requirements of private consumption (food, shelter, clothing, etc.); essential services of collective consumption (safe drinking water, sanitation, electricity, public transport, and health and education facilities); the participation of the people in decisions that affect them; the satisfaction of an absolute level of basic needs within a broader framework of basic human rights; and employment as both a means and an end (Friedmann 1992: 59–60). However, such aspirations clashed with the interests of Third-World élites who valued growth over redistribution, which greatly restricted the basic needs approach in practice.

Further innovations in development theory included recognition of the interaction of the economy with other aspects of society. Alternative development models gave primacy to non-economic considerations: 'socio-economic' and even 'social' replaced the conventional prefix of 'economic' in some circles, including the United Nations (D. M. Smith 1977: 217–25). A major international seminar on development ethics in 1986 concluded that an adequate definition of development must include the following: an economic component dealing with the creation of wealth and improved material conditions; a social ingredient of well-being in health, education, housing and employment; a political dimension embodying values such as human rights, political freedom, enfranchisement and some form of democracy; a cultural dimension recognising the fact that cultures confer dignity and self-worth; and a dimension called 'the full life paradigm', referring to the ultimate meanings of life and history as expressed in symbols and beliefs (Goulet 1990: 38). The main moral import was that there is much more to life than the fruits of economic growth, no matter how equitably distributed. Another example is the concept of 'responsible well-being', combining locally defined notions of well-being with personal responsibility, and incorporating well-being for all as the objective of development, livelihood security as basic to well-being, capabilities as means to livelihood, equity as prioritising the poor, weak, vulnerable and exploited, and sustainability – economic, social, institutional and environmental (Chambers 1997).

Despite all this, a review of the forty years covered by one of the development journals concludes that, in a very loose sense, development 'has always been about the politics of achieving social justice through economic change' (Harcourt 1997: 9). And David Simon (1998: 220–1) claims that 'the dominant aspirations of poor people and their governments remain concerned . . . with meeting basic needs, enhancing their living standards, and emulating advanced industrial countries in some variant of classic modernisation strategies'. Nevertheless, approaches which might claim universality have been targets for critique, refinement and reversal. The 1970s heard calls for a more critical assessment of modernisation along with an end to 'maldevelopment', for ecological prudence and for greater nurturance of civil society and local autonomy. To these clearly moral messages of their time, the 1980s added the demise of Keynsian economics and reconsideration of the welfare state. In the hands of the World Bank and the International Monetary Fund's structural adjustment programmes, neoliberal development strategies favoured a shift from public to private solutions and from state planning to the market, with distributive justice clearly subservient to economic growth. At the same time, there were calls for people's empowerment and participation, with women especially speaking out, and for respect of indigenous (non-Western) communities, cultures, identities and knowledge. This varied inheritance of the 1990s has encouraged a questioning of the entire notion of development based on state planning, with 'postdevelopment' added to the other lost certainties of modernity.

If anything remains certain in the field of development today, it is the significance of ethics. However, those aspects of development studies which most closely impinge on the discipline of geography have been slow to engage moral issues (D. M. Smith 1997c). A major exception is the attempt by John Friedmann to provide an alternative, 'morally informed' framework. He stresses 'the continuing struggle . . . for the moral claims of the disempowered poor against the existing hegemonic powers' (Friedmann 1992: 8), favouring the interests of those so often excluded both from design of the development process and from its benefits. His fundamental moral claim is that everyone is entitled to adequate material conditions of life and to be a politically active subject in their own community, based on the three foundations of human rights, citizen rights and the notion of human flourishing. The first two are central to the moral discourse of modernity, while the last one can be traced back to classical writings on the meaning of a good life.

In advancing his cause, Friedmann invokes the basic needs perspective, but with a relativist stance when it comes to implementation. Poor countries are unable to afford what might be considered basic needs in those that are rich, and there are also environmental considerations which makes winter heating more important in some parts of the world than others, for example, along with variations in people's tastes and preferences. He insists that a basic needs approach reflecting 'relative urgencies and scarcities' should be applied at the sub-national scale, by regions, towns and villages (Friedmann 1992: 63). A datum such as the number of calories per day needed for a healthy and active life has to be translated into a specific programme, which requires a territorially differentiated approach. Thus the spatial aspect of development encourages relativism at the level of what must actually be done. But Friedmann nevertheless espouses a universal conception of poverty as disempowerment, similar to the capability failure identified by Sen (1992).

Something of the diversity of contemporary geographical engagement with development issues is reflected in a special issue of *Third World Quarterly* (19[4], 1998; see especially Simon and Dodds 1998). As to the ethical dimension, Stuart Corbridge (1998) provides a review of the state of play, in an extension of earlier arguments (Corbridge 1992; 1993a; 1993b; 1994). He identifies a 'mainstream' perspective, which assumes that there is a broadly positive relationship between economic growth and human betterment, supposes that developed states have some responsibility and capacity for stimulating economic growth, and acknowledges that richer countries have a duty to poorer ones. He compares this with what he describes as the anti-intensionalism of the 'New Right', and with the anti-essentialism of a 'Post-Left'.

The New Right argues that inequality is not odious or wrong, that intentional-planned development is inefficient and undermines self-reliance and personal responsibility, and that government or international actions to promote development or equalisation in the short run can have damaging economic and social long-term costs. The Post-Left argue that developmentalism is a continuation of imperialism, reflecting the poststructural turn in social science (including postmodernism and postcolonialism), and insist on the heterogeneity and validity of different development experiences by commending a range of 'other' voices; their opposition to essentialism tends to undermine any strategy based on common characteristics of humankind. The New Right and Post-Left are forcing the mainstream to return to the ethical and moral foundations of development (Corbridge 1998: 41). Corbridge defends the intentionalist and internationalist views of an expanded mainstream development studies against both the New Right and Post-Left, stressing the claims that people in poorer countries might reasonably make on the resources of people in richer countries, on considerations of the ethics of development and transnational justice. A version of the place of good fortune (featured in Chapter 7) is implicit in his recognition of the significance of the chance involved in particular people being poor in poor regions and of their linkages with more affluent parts of the world, from which follows a minimal universalism with respect to the satisfaction of basic human needs (Corbridge 1998: 37).

It remains to mention some philosophical work which has exposed and expanded the ethical foundations of development studies. Prominent examples are the return to Aristotelian essentialism in reasserting human similarity (Nussbaum 1992), a Kantian emphasis on the duty of the rich to meet the basic needs of the poor (O'Neill 1986), the capabilities approach which continues to attract support as a source of universal transcultural norms (Sen 1992), and contributions with a feminist or gender orientation (e.g. Nussbaum and Glover 1995; Nussbaum 1998). Work on ethical dimensions of international relations (e.g. P. J. Anderson 1996; Graham 1997) and on justice among nations (e.g. Attfield and Wilkins 1992; Thompson 1992) adds to the diversity of the field.

To take the subject further, the remainder of this chapter adopts a case study of post-apartheid South Africa, to demonstrate something of what alternative development ethics mean, in the particular context of a distinctive society embarking on deliberate change which is intended to be for the good (based on D. M. Smith 1999c).

DEVELOPMENT AFTER APARTHEID

Apartheid in South Africa assailed the principle that people should not be advantaged or penalised for something over which they have no control, and from which they therefore derive no moral credit or fault. Race should not be a relevant difference, insofar as life chances are concerned, making institutionalised racial discrimination patently unjust. So, when Nelson Mandela committed his country's first democratically elected government 'to effectively address the problems of poverty and the gross inequality evident in almost all aspects of South African society' (RSA 1994: i), the hitherto disadvantaged black population could have expect to be the major beneficiaries. Indeed, the meaning of development might well have been taken as synonymous with social justice as racial equalisation.

Under apartheid the population of South Africa was classified into four 'race groups': Africans (about three-quarters of the national population of roughly 40 million today), whites (12 per cent), coloureds (8.5 per cent), and Asians or Indians (2.5 per cent). The magnitude of the task of racial equalisation is illustrated by the fact that two-thirds of African households live in poverty, by the criterion of the prevailing 'Minimum Level of Living'. Average monthly household income is over five times as high for whites as for Africans, with the Indians and coloureds also well below the whites. The difference between being born white or African is ten years of life expectation. The gaps are highlighted by a national desegregation of the United Nations Human Development Index (combining indicators of life expectancy, per capita income and education, so that the most developed countries have an index approaching one and the least developed closer to zero). South Africa as a whole scored 0.677, and was ranked 86th among 173 countries. 'White South Africa' scored 0.901 which would have placed it 19th, while 'black South Africa' (the African population) scored 0.500 and was ranked 119th. Scores for the Asians/Indians were 0.836, and for the coloureds 0.663 (O'Donovan 1995).

It is possible to envisage a staged process of development for racial equalisation, beginning with reallocation of resources, to promote equal capabilities, making equal opportunities a reality, and ultimately equalising living standards or well-being. The difficulty is illustrated by a calculation that, in round figures, to equalise per capita government expenditure on welfare services by former race groups, within the prevailing budget constraint, would require spending on whites to be reduced to one-third of what it was at the end of the apartheid era; to bring the other three groups up to white levels would require total spending to be trebled (van de Berg 1991). Projecting recent trends into the future suggests that it could take up to fifty years for racial shares of national income to equal shares of population. Thus, racial equality is not a realistic expectation within the lifetime of most present South Africans.

To focus exclusively on the racial dimension of inequality is flawed, however. Race is a social construct, the use of which was part of the process of domination and oppression under apartheid, and its deconstruction is implicit in the notion of a non-racial society. And there are also other dimensions of inequality to consider, notably gender. Furthermore, inequalities within the racial categories of apartheid are widening. In particular, class stratification of the African population is generating a fledgling middle class of independent entrepreneurs, a managerial aristocracy and a new political élite 'eager to join the consumerism of their former oppressors' (Adam 1994: 10).

Continuing or even widening inequality might be justified by its benefits to the worst-off, largely African population, according to Rawls's difference principle (see previous chapter). A common expectation in development strategy is that the poor gain from accelerated economic growth (the trickle-down effect), encouraged by production incentives manifest in income differentials. However, in practice the benefits may not reach the poorest sections of a society, resulting in greater polarisation between rich and poor and hence the possibility that South Africa is heading towards a form of 'socio-economic apartheid', in which the employed sector of the society progress but the numbers of unemployed or marginally employed increase relative to the ability of the state to provide survival benefits (Schlemmer and Levitz 1998: 77).

Recognition of the limitations of a conception of development as racial equalisation moves the discussion on to the strategies actually adopted by South Africa's postapartheid government, and the ethics involved. In the years of exile the African National Congress (ANC) was inspired by revolutionary socialism. The Freedom Charter, which provided the liberation movement's ideological foundation, included pledges to restore to the people the country's wealth, including minerals, banks and monopoly industry. Apartheid formally ended in 1994, when the ANC became the majority party in an interim government to hold power until the general election in 1999. A version of the ANC's major development strategy document, *The Reconstruction and Development Programme* (ANC 1994), or RDP, became official government policy (RSA 1994).

The RDP described itself as a coherent socio-economic policy framework, incorporating an integrated and sustainable programme, a people-driven process, peace and security for all, nation-building, linking reconstruction and development, and democratisation. These aspirations reflected the distinctive historical circumstances of South Africa's liberation from a system which had involved substantial state control, violent political struggle, national disunity and an absence of political rights for the majority of the population. An emphasis on empowerment meant the franchise for all South Africans, with attention also to enhancing the capacity of the civic associations and community organisations which had emerged in opposition to apartheid. The promise of fundamental transformation, involving non-racial democratic institutions and practices, along with a prosperous society embarking on sustainable and environmentally friendly growth, addressed 'the moral and ethical development of society' (RSA 1994: 4). And it did so in terms largely familiar to the progressive development discourse of the times, ambitious in scope but well short of revolutionary.

The first priority was to meet basic needs: 'jobs, land, housing, water, electricity, telecommunications, transport, a clean and healthy environment, nutrition, health care and social welfare' (ANC 1994: 7). This reflected not only South Africa's particular circumstances of massive poverty, but also a favoured strategy in international development circles (as explained above). There were some notable achievements, including 1.3 million homes connected to electricity and a million to water, a feeding scheme in 12 300 primary schools to combat malnutrition, and almost 300 new health clinics in rural areas (Marais 1998: 191).

The redirection of tax revenues into this kind of public expenditure to assist the poor clearly involved some redistribution at the expense of hitherto privileged people. However, meeting the ambitious objectives of the RDP depended more on the conventional association of equity with growth, recognising that poverty and inequality could be addressed only if the economy 'can be firmly placed on the path of high and sustainable growth' (RSA 1994: i). Merely keeping up with population increase requires economic growth of about 2 per cent a year, well above the average achievement of the 1980s. The slower the growth, the more the basic needs of the poor would require redistribution from the well-off, with its political risk of alienating sections of the (white) population whose acceptance of non-racial democracy did not extend to the erosion of their economic privilege.

The RDP had been promoted very much as a unifying strategy, almost beyond political debate. Cooperation rather than conflict between the state, labour and capital, as well as between the races, was the underlying assumption. All were expected to work together in a spirit of national unity, to enhance the economic performance from which elimination of poverty and racial equalisation would surely follow. However, there were ambiguities about the role and scope of the RDP, along with political and institutional weaknesses, which led to a failure to resolve the relative priority given to economic growth, reconstruction and redistribution (Blumenfeld 1997: 87–8). Rather than functioning as a developmental framework, it was criticised as an aggregation of policies 'designed to alleviate poverty without impacting on the complex of economic policies and practices that reproduce poverty and inequality' (Marais 1998: 92). While prioritising the needs of the poor, the programme failed to challenge the prevailing economic powers, and any residual socialist rhetoric was quickly replaced by 'economic realism'. Thus:

> The vision of a post-apartheid state developing creative and innovative alternatives to both soviet-style commandism and the uncontrolled free-market, to both east and west, is quickly fading as the emergent state succumbs to the power of money and organises itself along lines of capitalist rationality. The rhetoric that the RDP is 'people-driven' conceals the power of capital beyond apartheid.
>
> (R. Fine and G. van Wyk 1996: 22)

The provision of low-cost accommodation, for example, was left to the combination of individual initiative, meagre state subsidies and market forces which had so conspicuously failed to deliver in the terminal phase of apartheid; the state would not engage in housing production.

David Simon (1998: 241) sees the RDP as reflecting traditional modernist thinking, with the government as 'moral guardian' of reconstruction and development. However, popular disillusionment prompted by perceptions of limited achievements added to a growing uncertainty as to the state's commitment. Barely two years after the RDP's initiation, its national coordinating office was closed in 1996, signalling a shift in state policy. A new macroeconomic strategy, under the title of *Growth, Employment and Redistribution* (RSA 1996), or GEAR, emphasised reduced state spending, investment incentives, wage restraint, labour market flexibility and privatisation. This strongly resembled the neoliberal development strategy promoted elsewhere by the International Monetary Fund in its structural adjustment programmes (Adelzadeh 1996: 67), gestures towards which some commentators had found in the RDP itself (Adalzadeh and Padayachee 1994; Wolpe 1995). The priority given to employment creation reflected the truism that getting a job was the best way out of poverty. GEAR promised accelerated economic growth, to create 1.35 million new jobs by the year 2000.

The employment projections soon proved wildly optimistic, and GEAR attracted mounting critique. Not only had the macroeconomic policy 'failed dismally', so had sectoral strategies based on orthodox, market-friendly formulae, exemplified in particular by the failure of house-building to match that of the apartheid era (Bond 1998: 38). Furthermore, it was argued that the domination of neoliberal economic policy in the globalising international economy works against significant benefits for South Africa's poor, largely black majority: 'Navigating between the Scylla of international economic orthodoxy and the Charibdis of domestic political imperatives, the ANC-led government may well discover that it can deliver neither growth nor redistribution' (Price 1997: 173).

While much of the critique focused on technical aspects of the macroeconomic model generating the expectation of substantial job creation, Nicoli Nattrass (1996) showed how contrasting approaches to the labour market reflected different moral claims of contending positions. Thus big businesses supported GEAR in the implicit moral claim that lower wages are good for the poor, because a more flexible labour market encourages the expansion of relatively low-paid wage employment; so promoting the interests of capital through an investor-friendly environment is necessary for growth and promotes equity in the long run. The position of organised labour turns this moral argument on its head, proposing that high levels of inequality undermine growth, and that reducing inequality should be a precondition for economic growth rather than merely an outcome; so poverty should be addressed at least in part through improving the wages of low-paid workers. Her conclusion was that policy changes advocated on both sides 'are rooted in ideological assumptions and guesses about the functioning of the labour market and the determinants of investment' (Nattrass 1996: 39).

Thus we have, in effect, two different development ethics. In Rawlsian terms, the business and government position is that initial inequality in favour of the better-off can be justified as in the (eventual) interests of the worst-off. The opposition view is that initial moves towards equality are in the interests of the worst-off. What both positions share, at least at the level of rhetoric, is the imperative of benefiting the poor. If there are neither empirical nor theoretical means of adjudicating between these two alternatives (bearing in mind the uncertainty of the time frame in which benefits to the worst-off should be realised in a Rawlsian scenario), then the choice appears to involve incommensurate moral stands. This is the kind of impasse between competing metatheories to which postmodernism points, and which somehow has to be transcended if development as good change is to be identified and promoted.

The prospects for relieving poverty and for racial equalisation in South Africa do not look promising under either development strategy, however. If the aspiration of the impoverished black majority is, understandably, to achieve the living standards that they have seen the affluent (largely white) minority enjoy, and if this is what government is expected to deliver within a time span relevant to the present population, there will be massive disappointment. Hence the prediction of one critic:

> Left unchecked, the defining trends of the transition seem destined to shape a revised division of society, with the current order stabilized around, at best 30 per cent of the population. For the rest (overwhelmingly young, female and African) the best hope will be some trickle-down from a 'modernized' and 'normalized' new South Africa. This raises not only moral but political dilemmas, not the least of which is the danger that the incumbent élites come to view the excluded majority as a threat to newly acquired privilege and power, thereby introducing the spectre of a new bout of authoritarianism in response to social instability. (Marais 1998: 5)

Finding a way out of this dangerous trajectory, if there is one, raises broader questions of alternative development ethics, grounded in an alternative conception of the good.

ALTERNATIVE DEVELOPMENT ETHICS: A NEW SOUTH AFRICA?

At the root of the challenge of development and social justice in contemporary South Africa is the fact that the living standards which the minority were able to grant themselves under apartheid, and in the earlier colonial regime, were those of the Western European bourgeoisie. This is part of the postcolonial cultural legacy. As perceived by some black people, these remain the standards of the oppressor. And as well as being alien, they are unsustainable, in the sense of not being available to everyone in any realistic economic scenario under given resource constraints. A way of life which inevitably excludes a substantial majority of a society's population is hard to defend from a moral point of view. It is clearly inadequate to conceive of development merely in (re)distributive terms, racial or otherwise, in the context of the prevailing way of life of the affluent minority. An alternative conception of the good life is therefore required, to replace the possessive individual materialism which prevails among the well-to-do here, and elsewhere, in the contemporary world.

An opening is suggested by Augustine Shutte (1993). He tries to establish a position between the prioritisation of individual freedom associated with liberalism, in which the good life is a matter of personal preference, and the collectivism exemplified by socialism in which the good life is prescribed by the state. His central point of reference is the traditional African sentiment expressed by the Xhosa proverb '*umuntu gumuntu ngabantu*', or 'a person is a person through persons' (see Chapter 4): 'The aim of ethical and political life and thought is to discover, in the particular economic and geographical circumstances in which we live, the practices and institutions that best embody this ideal' (Shutte 1993: 90). In the same vein are calls for a movement 'that retrieves the communal ethics of African culture', referring to an organisation 'which has evolved its own philosophical system and its own way of interpreting and projecting reality . . . a communal structure which has affirmed its particularity' (Omo-Fadaka 1990: 176, 178).

But, on its own, the concept of community is inadequate. The reality of actual communities is that there may be a price for the strength of a common value system, in the form of suppression of difference and repression of minorities (as noted in Chapter 4). In South Africa under apartheid, a sense of community cohered around white identity, to the exclusion of others deemed different and inferior by virtue of skin pigmentation, strengthened by the geographical coincidence of residence and race group. 'Different moral communities, delineated by racial, linguistic, class and other divisions, have co-existed in uncomfortable proximity' (Atkinson 1991: 126), creating a distinctive moral geography encouraging antipathy towards difference. So, a way has to be found of enabling the strength of communitarian sentiments of mutuality to be augmented by respect for, and inclusion of, different others.

An important feature of post-apartheid political discourse is a search for perspectives grounded in Africa and Africanism. Both Nelson Mandela and his successor as state president, Thabo Mbeki, have invoked the 'African Renaissance' which has featured in some other countries. It has been summarised as follows:

> It is the rebirth of the African continent after centuries of subjugation. It is about the redress of knowledge, of correcting negative images inculcated into its people . . . about seeking new ways of thinking and feeling about Africa, its history, and its

> economic, social and political status. It is an invitation to re-invent ourselves and what we do, how we do it and who benefits from it. (Ntuli 1998: 16–17)

Central to the African Renaissance is the notion of *ubuntu*, which has attracted interest in circles seeking an alternative development ethics in the contemporary spirit of authenticity of indigenous culture.

At its simplest, *ubuntu* is 'a metaphor that describes the significance of group solidarity' (Mbigi and Maree 1995: 1), a widely held value of which is to prolong the lives of all members of the group (Silberbauer 1991: 18). A practical example is provided by Gerhardus Oosthuizen (1996), in a study of African independent or indigenous churches (AICs). As a grass-roots mobilisation of the poor, the AIC movement is probably the largest institution of civil society in South Africa, with potential for involvement in social and economic development issues. The AICs can take the form of 'geographical communities in small villages [which] replicate in the anomic contemporary South African environment the traditional black rural social order' (Oosthuizen 1996: 310). Central to this is emphasis on membership of a community:

> This emphasis is one which runs deep in traditional African society. *Ubuntu* is the name given to the ethos of mutual support whereby individualism is harmonized with social responsibility, which underlies the fabric of traditional African society. But in the AICs, *ubuntu* finds a complement in Christianity: not the secularized Christianity which has become a feature of the developed world, but the mutualism and communal sharing which was a characteristic of the early Christian church . . . The basis of *ubuntu*, the extended family network wherein individuals found self-expression through their support of the rest of the household, has been severely eroded by colonialism and apartheid . . . the extended family survives as a template for the AICs: small congregations under the authority of a respected leader, whose members offer each other support in times of crisis. (Oosthuizen 1996: 312–13)

The social action and compassionate support of the AICs has helped to address material and psychological consequences of apartheid, including the common African experience of migration to the city. They create 'a caring enclave in the impersonal urban world', or a 'psychic zone' within which the uprooted and disoriented individual can adapt (Oosthuizen 1996: 315). Another writer describes how someone in need created 'a web of solidarities with women co-congregationalists, a community of care that gathered every Sunday at the church' (Sitas 1995: 85).

The incorporation of a spatially extensive ethic of care into the kind of communitarianism invoked by the concept of *ubuntu* would go to the heart of the problem of responsibility to others, who may be distant as well as different. Some feminists note the affinity of an ethic of care with strands of African-American religious traditions which reflect an 'Afrocentric morality' (Harding 1987: 364; Tronto 1993: 83–4; Hekman 1995: 76,

107–8). Thus, there is the possibility that aspects of both contemporary feminist and traditional African thought might converge on a conception of the good which transcends possessive individualism in favour of a relational self prioritising responsibility to care for others, and especially those in most need. This might lead to a contextual development ethics grounded in South Africa, but with wider aspirations, if it can incorporate different others into the relevant moral community.

However, there are doubts as to the capacity of the notion of *ubuntu* to do the work required for the creation of a broader national community in South Africa. Some see *ubuntu* as merely a benign, reified and romanticised version of the rural African community (Wilson 1996: 11), an idealisation which may obscure the actual parochialism and patriarchy of the traditional community. And there is the 'dark side' of *ubuntu* to recognise: 'Individual conformity and loyalty to the group is demanded and expected . . . failure to do so will mean harsh punitive measures' (Mbigi and Maree 1995: 58).

There is also the burden of geography and history to overcome. This includes the divisive impact of apartheid in general, and such particular manifestations as the migrant labour system which divided families. 'The roles of traditional stabilizing and unifying social institutions like family structures, churches or community bodies have declined or disintegrated, leaving the country's tattered social and moral fabric largely unattended' (Marais 1998: 110). Basic elements of social cohesion are therefore lacking.

A conscious effort has been made in post-apartheid South Africa to come to terms with the past, and to use it creatively in the formation of a new, united and harmonious community on a national scale. This nation-building project is reflected in the idea of the 'Rainbow Nation', in which population diversity is viewed as a positive asset, in the spirit of multiculturalism, but within a broader affiliation with the 'New South Africa'. From the time of its initial election victory the ANC leadership projected a non-racial definition of South African identity, as a reflection of their own normative vision (Price 1997: 168). There is increasing trans-racial interaction. And there are growing racially-mixed residential areas, where 'the romance of the "New Nation" is being attached to the notion of integration to generate a relevant political meaning for these previously transgressive spaces' (Robinson 1997: 380), in which racial separation was challenged under apartheid.

New South African citizenship is therefore supposed to transcend ethnicity and race, in a changing politics of identity. The task is to strike a balance between nation-building and the rights of citizens to choose from the menu of identities available in a culturally diverse society (Bekker

1997: 17). An important instrument in the post-apartheid construction of citizenship is a Bill of Rights, incorporating both political rights and a generous array of socio-economic rights. The discourse of human rights brings together diverse political constituencies, including blacks previously denied citizenship rights and whites reassured by commitment to an individualistic and liberal political economy; it also contributes to a mode of state-building where citizenship coheres not to the nation, with its ethnic or racial overtones in South Africa, but to the state, in an attempt to 'construct overarching unity metaphors and to cement collective moralities' (Wilson 1996: 10).

However, inscribing rights in a constitution is not the same as guaranteeing them in practice. Welfare rights, with their egalitarian implications, are more difficult to implement than liberty rights, in the prevailing environment of liberalism. Welfare rights also require substantial state resources, which may not be forthcoming. In the present circumstances it is hard to see what the statutory right of access to housing, water, social security and a healthy environment actually means, and within what time scale, to the seven million or so South Africans living in informal 'shack' settlements. If, as seems inevitable, the Bill of Rights makes little difference to the material conditions of the poor, how will they experience the benefits of citizenship within a supposedly united state, still deeply divided by class if less so by race? Problems can be expected when issues come before the Constitutional Court:

> How will it resolve the tension between the competing values of freedom and equality? How will it, in addition, give meaning to each of these diverse concepts? Will freedom be construed as connoting that the rights of individuals must take precedence over the rights of groups, or vice versa? Will the right to equality be understood as requiring equality before the law, equality of opportunity, or equality of outcomes? (Jeffery 1997: 18)

Resolving such moral questions in the particular context of post-apartheid South Africa is a crucial part of the practice of development.

Another important initiative has been the Truth and Reconciliation Commission (TRC). This followed precedents in a number of other countries coming to terms with acts like torture and murder committed by oppressive regimes (Botman and Peterson 1996; Boraine, Levy and Scheffer 1997). The guiding assumption was that finding out the truth of what happened would assist reconciliation and nation-building. The Commission's hearings, chaired by Archbishop Desmond Tutu, involved harrowing evidence of the fate of victims, along with confessions by perpetrators in return for amnesty: 'a novel and controversial experiment in public political morality' (du Toit 1997: 7). Truth and reconciliation became 'ubiquitous moral metaphors of a revised national identity'

(Wilson 1996: 14). However, the critical reaction to a critical review of the TRC (Jeffery 1999) suggests that it may have been more divisive than unifying. There is also the problem that much (perhaps most) harm was done under apartheid by systemic means, perpetuating structural inequality and discrimination, for which it is hard to attribute individual blame. If the truth about events is still contested, moral truth remained even more elusive.

Thus there are questions as to the effectiveness of these and other initiatives, in affirmative action for example, in helping to create a cohesive national community. There are also powerful impediments to the development of a more egalitarian society which prioritises care for those in greatest need, in the contemporary neoliberal ideology of reverence for markets. The impetus to privatise state industry and services carries special temptation in the new South Africa, to which the ANC has yielded with alacrity. The circumstances in which such activities were run under apartheid suggests that there could be gross inefficiencies, and, if services are to be extended to those sections of the population previously discriminated against, private capital could relieve state expenditure with its many competing demands. However, privatisation will lead to further consolidation of wealth and power in the hands of those conglomerates capable of purchasing public corporations (B. Fine 1995: 20). There is also the need to recognise the role of public goods with positive externalities such as preventing the spread of disease (Bond 1998: 39), otherwise privatisation could frustrate rather than facilitate national development efforts favouring the hitherto disadvantaged.

South Africa's experience of different development paradigms clearly reflects its difference, or particularity. Under apartheid the government attempted a process of territorial disagregation, assigning African tribal groups to so-called homelands in which they were expected to exercise their political aspirations as independent nations, leaving the vast majority of the original state for the whites (along with the Indian and coloured populations, in a subservient position). And this was despite the fact that about half the African population lived in areas designated as 'white', mainly the cities where they worked. The prevailing mainstream discourse of development planning was adopted, including industrial dispersal from the metropolitan cores to peripheral growth points (Fair 1982). While this might have given the impression that South Africa was conforming to some universal, technical and political neutral conception of development, the purpose was particular to that situation at that time: to facilitate 'influx control', which meant restricting black residence in the cities. When apartheid eventually came to an end, it was largely because the spatial grand design could not be sustained; it attempted the impossible – to separate the capacity to labour from its human embodiment. White South Africa wanted black labour, as cheap as possible, but without the rights associated with being human rather than an inanimate factor of production.

The demise of apartheid coincided with that of the socialism which had been the major inspiration for the liberation movement in exile. Once the ANC came to power, their major innovation in development policy was an emphasis on basic needs within an integrated national strategy, which was itself soon to yield to the neoliberalism rampant elsewhere. South Africa's tragedy is to be faced with such an inheritance of inequality and poverty, in this era of globalisation, and of associated shifts in the ideological, political and economic spheres away from state involvement in the development process. As one prominent critic sees it: 'Neo-liberalism represents a political/ideological effort to ride the tide of globalization in the best possible interests of the most powerful sets of national economic élites' (Marais 1998: 115). Its ascendancy reflects the current failure of an alternative ethics to come to terms with globalisation.

It is to economic growth that South Africa now looks for the resources to provide for basic needs, yet it is in this respect that post-apartheid society is particularly constrained. Internal capital accumulation is unlikely to generate the resources to raise economic growth substantially above the level required to keep up with population, and investment from overseas depends crucially on how international capital perceives South Africa's comparative advantage and political stability. Vigorous redistribution at the expense of the affluent minority would feed back negatively to business confidence. The recognition that more radical strategies of societal transformation could seriously prejudice economic growth underlines how much the government's freedom of action is limited. Thus: 'at the very moment when democratisation stimulated the popular demand for better social and welfare services, structural adjustment required that they be denied' (R. Cornwell 1998: 12). And the mood of anti-developmentalism or postdevelopmentalism among what Corbridge (1998) identifies as the 'Post-Left' hardly encourages grand projects initiated by the state.

The changes in ethics required to define the good life in terms of prioritising care for those in need are themselves constrained by the prevailing liberal ideology supportive of possessive individualism. This is part of the broader culture in which ever more aspects of life are incorporated into the commodification on which capitalism in its present form seems to depend. The challenge for development is somehow to break these links of interdependence. In considering this challenge, it is important to recognise the historical depth as well as the moral strength of the connection between communitarian thinking and egalitarianism. From

Rousseau's conception of a democratic community to contemporary communitarianism, there has been an understanding that group harmony, mutuality and collective responsibility may be threatened by substantial material inequalities among members. Some altruism may prevail in a highly unequal society, but it is more likely if people live similar enough lives to recognise their similar needs and shared humanity.

For any nation to act as a community is difficult enough. To underwrite a Rawlsian or egalitarian conception of justice, or an ethic of care, people need a collective identity, 'the kind of social bonds which incline them to accept a duty to promote the welfare of their community and everyone in it' (Thompson 1992: 104). In South Africa there are special impediments imposed by a past in which groups of people defined by race and tribe were encouraged to think of themselves as distinct nations. Apartheid fed this kind of identity and a strong sense of others as different. A feeling of national unity as New South Africans might attain sufficient strength and purpose to transcend inherited divisions and indeed to challenge liberal individualism in favour of a collective responsibility for the poor. But to create and sustain this is a massive task, even in a climate of reconciliation, in a country and a world where contrary values and forces predominate. Hence the voice of economic realism: 'Capital is all-powerful . . . While international capital flows may not yet hold the key to a country's economic life or death, policy makers ignore this power at their peril' (Nattrass 1996: 34, 39). The power of capital is also reflected in the priorities of the prevailing neoliberal development discourse (Figure 8.1).

We close with a reminder of the material reality of life in the New South Africa from Ashwin Desai, a sociologist whose regular newspaper column sustains a provocative critique:

> We live in a society where millions of people are condemned to extreme deprivation even before they are born. Regardless of their own qualities, they are condemned to misery and humiliation. A few others are born into utter luxury, and waited on hand and foot for the rest of their lives. Ours is a world where it is natural for some inhabitants to eat R500 [£50] meals while others, literally metres away, cannot afford even a loaf of bread. (Desai 1999: 63)

He goes on to bemoan the normality of giant corporations shedding loyal workers, which is the experience of a growing number of South Africans, adding to pressure to reduce public sector employment in a country desperate for jobs. He concludes that one cannot really comprehend the true evil of it all, centred on the global worship of money which, as Marx predicted, 'becomes the real community, since it is the general substance of survival for all' (Desai 1999: 64).

The evil of apartheid was transparent. Reverence for Mandela as moral saint has helped to obscure the deeper evil with which the 'New' South

Figure 8.1 *The moral dilemma of intergenerational justice posed by the development strategy promoted by the International Monetary Fund (IMF), requiring present sacrifices for supposed future benefits. (Source:* Environment and Urbanization, *10 (1), April 1998, reproduced by permission of Paul Fitzgerald)*

Africa now has to struggle. To challenge this requires a development ethics in which those with more wealth and power not just accept having less, but welcome this as a means to a better quality of life, as this chapter's epigraph suggests. The prospects are not encouraging.

CHAPTER 9

NATURE: ENVIRONMENTAL ETHICS

> [E]nvironmental ethics potentially addresses some of the most profound questions that confront late modern societies, whose widespread level of concern over environmental crisis spanning local to global proportions is indisputable . . . any normative environmental ethic must be built on a differentiated sense of culture and the differential responsibility of various human actors for environmental well-being.
>
> (Proctor 1998d: 235, 246)

From the perspective of human geography which has preoccupied this book so far, nature or the physical environment constitute the 'other' side of the discipline. Such has been the divide, over the past three decades, that human geographers have regarded the physical largely as a separate sphere of inquiry. This has not always been the case: in the discipline's formative years its primary focus was man–environment relations, with the subject matter often arranged in the sequence of 'physical background' and 'human response'. Now, there is renewed recognition of the interdependence of the human and the physical, prompted in part by the perception of an impending ecological crisis which, at its most apocalyptic, threatens the very survival of life on Earth.

This is all part of the changing understanding of the links between humankind and environment. As to nature, this is now commonly regarded as a social construct rather than a fixed and universal scientific category, the meaning of which may differ among cultures as well as changing over time (Harvey 1996; Gerber 1997; Proctor 1998c). David Livingstone (1995) recognises that the natural, the cultural and the moral have always been intimately related, indeed mutually constituted. Further:

> The very concept of environmental problems presupposes some normative state of nature. To speak of an ecological problem is to make an ethical judgement that society would be better off without it. For this reason the environmental debate is frequently, at base, a debate about what constitutes the good life.
>
> (Livingstone 1995: 370)

He identifies three different ways of representing nature, and shows how these have influenced moral stances towards the environment. The first is the notion of an organic world, supported by the image of Mother Nature, and more recently by ideas concerning the partnership of all forms of life as valuable in themselves. This induces an attitude of reverence or piety towards nature. The second is a mechanistic vision, of an environment requiring management. This tends to induce the value of utility or the wise use of nature, guided by insights of science and commitment to social progress: conservation in the interests of humankind as opposed to preservation in some pristine state. The third is an ecological conception of nature, as a self-sufficient, coherent, stable, integrated entity. Community rather than piety or utility is the key virtue of this 'ecosocial economy', in which the human and nonhuman are intimately interconnected.

It is in this context of different understandings of nature that the environment has become an important focus of normative thinking in recent years. The context also includes the emergence of 'green' activism and politics, in response to growing awareness of the dangers of environmental pollution, the limitations of non-renewable resources and the possible hazards of biotechnology in such forms as genetically modified crops. The commercial applications of new technology often appear to be well ahead of scientific understanding of possible consequences, and even more advanced than consideration of ethical implications.

Geographical engagement with the interdisciplinary field of environmental ethics is building new bridges over the discipline's human/physical divide. Recent books include those of Pepper (1993), Harvey (1996) and Low and Gleeson (1998), and collections edited by Peet and Watts (1996) and Light and Smith (1997). In all these publications, the perspective is very much that of human geography reaching out to incorporate the environment, and so it is with the treatment here. This chapter provides an introduction to environmental ethics, followed by brief discussion of specific topics selected to provide links with themes elaborated in previous chapters.

INTRODUCING ENVIRONMENTAL ETHICS

Like development ethics, with which it has close connections, environmental ethics emerged as a distinct field of inquiry in the late 1960s and early 1970s. By the end of the twentieth century the number of books linking environment and ethics had become too numerous to cite, yet there is no single text which portrays convergence on clearly specified content. Nor is there unanimity as to the way the field is defined, at anything other than a general level. For example, environmental ethics, or environmental philosophy, is described by Booth (1997: 256) as 'an exploration of the cosmos and humanity's relationship to it . . . the marriage of ecology and philosophy'. Others are more specific: Proctor (1998d: 236) cites the following definitions of environmental ethics in two anthologies: 'the field of inquiry that addresses the ethical responsibilities of human beings for the natural environment', concerning itself with 'humanity's relationship to the environment, its understanding of and responsibility to nature, and its obligations to leave some of nature's resources to posterity'. Light (1996: 275) defines environmental philosophy as 'the attempt to bring the traditions, history, and skills of philosophy to bear on the question of how to maintain the long-term sustainability of life on this planet'. The references to posterity and sustainability signal the primary concern with human survival. The term 'global ethics' is sometimes deployed to invoke the interdependent world of environmental problems.

Unlike development ethics, environmental ethics has generated a substantial literature of a primarily philosophical orientation. A further definition is coupled with a summary of the problems encountered, as follows:

> Environmental philosophy engages in critical reflection on the moral rules, conventions, and shared meanings that inform human inquiry and our relationships to nature. As a relatively new branch of moral philosophy, environmental ethics arises out of increasing unease with the destructive consequences of human economic, technological, and cultural practices. Many philosophers are unpersuaded that reliance on traditional utilitarian and rights-based deontological moral theories can adequately respond to the problems, both philosophical and practical, that need to be addressed. There is, therefore, a growing absence of consensus about the underlying moral framework that should define and structure the practices of knowing, valuing, and using nonhuman nature. (King 1997: 209)

There is a debate between monists, who seek the conviction of a single moral theory for environmental ethics, and pluralists, who point to the heterogeneous sources of value in nature and to the multitude of contexts in arguing for a diversity of approaches. A review of the work of two influential writers makes the point that the monist J. Baird Callicott (1994) nevertheless understands how a variety of world cultural traditions can inform environmental ethics while the pluralist Arne Naess (1989) argues that the foundations of Christianity, Buddhism and some schools of philosophy all provide support for 'deep' ecology (Light 1996: 288–9). The problem of relativism is recognised, yet the bulk of work in environmental ethics has not yet engaged the implications of multiculturalism (Proctor 1998d: 249).

Questions of value are fundamental in extending the realm of ethics to the nonhuman world. While the intrinsic moral worth of all members of humankind has been recognised since Kant, dealing with the environment underlines the difference between the purely instrumental value of nature to humankind and its possible intrinsic value. This is the distinction between anthopocentrism and various nonanthropocentric perspectives which may be referred to collectively as physiocentrism (Proctor 1998d: 246). The former privileges human preferences, interests and needs, and weighs harm to nonhumans or disruption of ecological systems only insofar as these constitute a cost or benefit to human beings (King 1997: 210). The latter might be considered a defining characteristic of environmental ethics itself: 'positions which assert that at least some nonhuman organic entities ought to be valued for reasons not reducible to their use value' (Cuomo 1997: 12). While anthropocentrism stresses human distinctness from nature, physiocentrism incorporates nature into the moral community. Some, including deep ecologists, see no ontological divide between humans and nonhumans: humans are embedded 'in constitutive relations with the nonhuman world' (Whatmore 1997: 45).

The practice sometimes referred to as ethical extensionism raises the question of the relevant criteria for defining the boundaries of moral considerability or standing. Moving from a human-centred ethics to one which is animal-centred or life-centred, or to some form of ecological holism focused on large ecosystems or on the biosphere as a whole, involves abandoning the criterion of capacity to reason, as this is confined to humans. Invoking the capacity for pain includes animals, but not plants. All living things might be seen as subjects of intrinsic value as forms of life, or as possible expressions of beauty – a reminder of links between aesthetics and ethics. A case is sometimes argued for biological egalitarianism, with all living things assigned the same moral value. However, extensionism can allow differences in moral significance, such that all forms of life have moral standing but some (usually humans) are accorded greater significance than the rest.

The treatment of animals reveals some of the difficulties of nonanthropological extensionism. There are a range of ethical issues, including whether animals should be raised and killed for food, tormented for entertainment, used for scientific experiments, or bred for human spare parts. The capacity for pain and suffering has become widely accepted as a criterion for the moral considerability of animals, but the major ethical traditions of rights and utility both encounter problems when extended in this direction. Animals may be accorded a right to life and to respect, but not to vote or to religious freedom. And they cannot be expected to recognise other rights, such as for lions to respect the life of impala. 'Nature knows no rights . . . is amoral . . . animal rights do not exist in the absence of persons' (Rolston 1993: 256–8); any animal rights are social constructs, with strong elements of cultural specificity. As for utilitarianism, such celebrated cases as whether to eject a human or a dog to keep a lifeboat afloat defy simple resolution, if the sum total of happiness of lives saved (weighted equally) could be maximised by disposing of a miserable human rather than a contented dog, or if the preferences of humans for their own species enter the calculus (Gruen 1991). Replacing the reason underpinning mainstream ethical theories by sympathy could prioritise humans over animals, or cats over cows, or even pet animals over some humans. A robust animal-centred ethics may be more convincingly grounded in a relational ethics recognising structures of power and oppression than in liberal individualism (Cuomo 1998: 93–8, 101–5).

Extending an anthropocentric ethics into the natural environment itself produces problems. A human right to appropriate the wherewithal to live from the natural environment is entailed by a human right to life:

> If there is such a thing as natural rights, privileges we possess just by being born on Earth, the right to nature is foremost among them . . . a right to the natural environment includes protection of air, soils, waters, essential biological processes, the sustainable productivity of the land, preservation of biodiversity, protection against contamination by toxic substances, access to natural resources essential to life, and perhaps access to public lands and commons. (Rolston 1993: 260)

While such environmental rights have been recent additions to the more familiar human rights, they are attracting increasing attention. For example, the right to an environment not harmful to health or well-being in the South African constitution of 1996 includes a right to environmental protection, to prevent pollution and ecological degradation and to promote conservation and ecologically sustainable development (Jeffrey 1997: 121). However, such rights may be extremely difficult to secure, especially in a world of private ownership and control of most resources. The distribution of disbenefits of development, in the form of environmental pollution, adds to (or subtracts from) the more familiar benefits of enhanced living standards: the problem of environmental equity or justice, taken up in the next section.

As cultures intertwine with their landscapes, they may require conservation of certain natural goods (Rolston 1993: 260–3). To the bald eagle symbolising American self-images and aspirations of freedom, strength and beauty may be added the white cliffs of Dover symbolising English insularity from continental Europe, the sacred sites of Australian aboriginal people, and so on. In a different context, the coniferous forests of the Pacific north-west of the United States and their spotted owl symbolise to 'environmentalists' the values of receding wilderness in the face of commercial resource exploitation (Proctor 1995; 1999). If humans have a right to liberty and the pursuit of happiness as well as to life, then such natural features can be significant reflections and repositories of value. However, moves for their preservation are not easily promoted within the paradigms of individual rights or utility.

The same can be said of sustainable development (discussed later in this chapter). Preoccupied with the conservation of nature for human use, this perspective remains anthropocentric, yet if it is to be effective it requires a more comprehensive environmental ethics; respecting the ecosystesms on which sustainability depends requires naturalistic as well as humanistic reasons (Rolston 1993: 266). Hence the moves in recent years towards nonanthropocentrism as a holistic biocentric project, 'locating a form of intrinsic moral value in nonhuman nature, independent of the interests, preferences, and rights of human beings' (King 1997: 215).

The usual starting point goes back to the early years of the American conservation movement. Aldo Leopold (1949) sought an ethic capable of meeting situations so new or intricate, or involving such deferred reactions, as to require modes of guidance grounded in community rather than in individuals. He explained what he termed the 'land ethic' as follows:

> All ethics so far evolved rest upon a single premise: that the individual is a member of a community of interdependent parts . . . The land ethic simply enlarges the boundary of the community to include soils, waters, plants, and animals, or collectively: the land . . . a land ethic changes the role of *Homo sapiens* from conqueror of the land-community to plain member and citizen of it. It implies respect for his fellow-members, and also respect for the community as such. (Leopold 1949: 239, 240)

The ecosystem or the entire biosphere may be understood as having intrinsic value, with individual members of instrumental value in its maintenance. As in the conventional human community, individuals also have intrinsic value, but as members of the wider whole rather than as isolated entities. This conception, subsequently associated with the deep ecology movement, compares with anthropocentrism in which humans have intrinsic value but other species only instrumental value, and with biocentric individualism in which individual animals and plants are the locus of intrinsic value.

Biocentric or ecological holism addresses some of the limitations of a purely anthropocentric environmental ethics. But as a case of the historical-geographical specificity of ethics, its origins in American preoccupation with the preservation of wilderness may overvalue one particular human interests at the expense of others, like farming the land for food to prevent human starvation. Deep ecology has been criticised as similar to communism, fascism and even Nazism, in advocating that the interests of some collective (like the ecosystem) should override those of the individual (Stein and Harper 1996: 97). It appears to extend maximal protection to wild nature while devaluing human culture and individual moral worth. One critic concludes:

> While biocentrism plays an important role in compelling us to rethink the contemporary compulsion for economic development, wasteful production, and ignorant expansion, it provides little positive guidance on how humans might live harmoniously and legitimately within a nature that contains both human and nonhuman aspects. (King 1997: 222)

Reactions include the call for a more contextual environmental ethics, taking seriously the concept of place as grounding the practice of moral inquiry and situating moral and social action (King 1997). Approaches prioritising context, diversity and difference include bioregionalism, which invokes a traditional geographical concept of the region as 'a piece of land that is defined by physical, biological, and cultural characteristics' (Booth 1997: 258). This has the benefit of recognising local cultural understandings, providing that respect for the environmental wisdom of indigenous peoples is not exaggerated. William Lynn (1998: 231–8) argues for an earth-centred 'geocentrism', with individuals, species and ecosystems having concurrent moral value as intrinsic ends in themselves as well as instrumental means to other ends; who (or what) has moral standing and which moral values are relevant should be more place-sensitive than under biocentrism or ecocentrism. Recognition of the significance of difference has been consolidated by David Harvey (1996), who elaborates the interdependence of the socio-political and the ecological in particular contexts, including a reminder of the often neglected urban dimension of environmental problems.

This brief introduction to major themes in environmental ethics has of necessity glossed over many details and refinements of argument. Some of these will emerge in the discussions of the particular issues of environmental equity or justice, sustainable development and the role of community and care, which occupy the rest of this chapter.

ENVIRONMENTAL EQUITY AND JUSTICE

A criticism of environmental ethics is that preoccupation with philosophical issues can be at the expense of engagement with actual environmental problems, and with the people who experience them. The concept of environmental equity or justice is helpful in providing something more tangible than broader ethical concepts such as value (Almond 1995: 4). It has attracted considerable geographical interest (Harvey 1996: Chapter 13; Heiman 1996; Low and Gleeson 1998:

Chapter 5), compared with the limited attention given previously to facility location and its environmental impacts (D. M. Smith 1977: 329–35; 1981: 366–78). The origins of the environmental justice movement are so closely associated with the United States as to prompt a British commentator to ask whether this American trend may 'cross the pond' (Walker 1998: 358).

In 1990 the United States Environmental Protection Agency established an Environmental Equity Workshop, which partially reaffirmed earlier studies finding a strong correlation between the location of commercial hazardous waste facilities and the percentage of 'minority' residents in a community. In 1994 President Clinton signed an executive order requiring every federal agency to achieve the principle of 'environmental justice' by ameliorating the impact of their activities on low-income populations, assumed to be racial minorities. International interest was signalled by affirmation of a healthy environment as a basic right for all the Earth's inhabitants by the United Nations in the 1992 Rio Declaration, though this has no greater practical force than other such proclamations of universal human rights.

Susan Cutter has identified different though related dimensions of the issues involved. 'Environmental equity is a broad term that is used to describe the disproportionate effects of environmental degradation on people and places' (Cutter 1995: 112). She makes a distinction between process equity, concerned with causal mechanisms such as the role of social and economic factors in creating differentiated landscapes of risk (or 'riskscapes'), and outcome equity concerned with the spatial-temporal distributions of benefits and burdens. 'Environmental justice is a more politically charged term, one that connotes some remedial action to correct an injustice imposed on a specific group of people', such as protection from environmental degradation, prevention of adverse health impacts, mechanisms for assigning culpability and redress of impacts.

The environmental equity or justice movement represents a shift in spatial focus from the 'white upper-class environmental rhetoric' surrounding preservation of distant pristine habitats to 'environmental improvements in the quality of life closer to the homes of affected residents' (Cutter 1995: 113). In a field that proliferates acronyms, it reflects a move from NIMBY ('not in my back yard') towards NIABY ('not in anyone's back yard'), BANANA ('build absolutely nothing anywhere near anybody'), or even NOPE ('not on planet Earth'), with respect to LULUs ('locally unwanted land uses').

An issue crucial to environmental equity is captured in another acronym: PIBBY, or 'put it in black back yards'. This is usually referred to as environmental racism (Low and Gleeson 1998: 107–10; Sterba 1998:

140–3), reflecting deliberation in the disposal of waste and the siting of noxious facilities in the United States close to black residential areas, observed most famously as 'dumping in Dixie' (Bullard 1990). As Cutter (1995: 118) observes, 'there is substantial evidence that people of colour in the USA bear a disproportionate burden of environmental hazards'. She suspects that this is also true in other nations. That the relevant criterion of vulnerability may be class rather than race is central to debates on environmental equity and justice:

> Hazardous wastes, together with the factories and power-stations which produce them, are more likely to be found in the neighbourhoods of the poor than those of the rich; the prevalence of diseases caused by air pollution will be greater for the poor than for the rich; the costs of living in an environment with inadequate recreational facilities will be proportionally greater for the poor than for the rich.
> (Gower 1995: 50)

However, such associations are easier to assert than to demonstrate. This is in part due to familiar technical difficulties in geographical research, involving the identification of the appropriate scale at which to measure correlation between pollution and population characteristics, and to the problem of whether it was the source of pollution or the vulnerable population which came first. Thus, empirical investigations in the south-eastern United States conclude that the story of environmental justice is 'more complicated than simple correlations between race, income, and toxic exposure . . . racism seems to account for only part of the process of the social production of risk' (Cutter and Solecki 1996: 395).

As in the field of social justice, a purely distributional conception of environmental justice is open to criticism. For example, Robert Lake (1996: 160) accepts the principle that 'environmental problems should not be disproportionately concentrated in poor communities or communities of color or, more broadly, in socially, economically or politically disempowered communities', but points to situations in which this might conflict with equally compelling principles of local autonomy and self-determination. One such might be the case of 'volunteerism', under which a poor community accepts the location of a waste dump, for example, for the sake of its jobs and tax revenues. Application of the general equity principle under which the dump might be located elsewhere ignores local consequences, or the contextuality of environmental justice.

The critique goes further, to recognise the importance of the broader institutional and structural context, including the processes of domination and oppression which assign some persons to vulnerable places. This requires attention to the decision-making process. Production decisions linked to cultures of consumption generate the environmental problems which become the subject of distributional conflict, with (local) politics tending to discriminate against minorities and the poor. Achieving environmental justice requires not simply a redistribution of those problems, but a change in the process that leads to their production: the unequal power relations of capitalist societies (Lake 1996: 169). The advice to geographers is clear:

> Environmental inequity will not be demonstrated through a mapping of distributions but rather through a mapping of power relations that link communities to the institutional structures creating the burdens to be distributed . . . an assessment of the processes creating landscapes within which distributions are embedded . . . entails a focus on the structure of production and on the ways in which communities are linked (or not) to capital investment decisions and to the process of uneven development. (Lake 1996: 170)

The perspective required recognises the interdependence of environmental and social justice, in theory and in practice. The local and particular manifestations of environmental inequity have therefore to be linked back to broader social processes and schemes of ethics. The objective is 'a discourse of universality and generality that unites the emancipatory quest for social justice with a strong recognition that social justice is impossible without environmental justice (and vice versa)' (Harvey 1996: 400). Both are constitutive of the good life.

The focus of much activism and debate in environmental equity and justice is at the local level. This is understandable, in view of the predominantly local scale of both experience and regulation of pollution. But there are other scales to consider: international, intergenerational and interspecies (Cooper and Palmer 1995). The international scale involves such problems as the export of environmental hazard, not only in the form of the (re)location of noxious facilities in poor rather than rich countries, but also in spillover effects such as acid rain and the fallout from disasters, like the Chernobyl nuclear power station, the spatial scope of which transcends national boundaries and their sovereign jurisdictions (Kuehls 1996). The intergenerational scale includes bequeathing a polluted environment to the future, along with the storage of dangerous waste out of sight (and mind) until the containers begin to leak. The interspecies scale, or 'ecological justice' (Low and Gleeson 1998), involves humans passing environmental costs onto others, for example animals displaced when forests are cut down or fish poisoned in a polluted river.

There is interdependence among these different scales. Success of the environmental movement in developed countries may accelerate the relocation of hazardous industries elsewhere (Pulido 1996; Low and Gleeson 1998: 122), for example across the United States border into Mexico, so that justice within a nation may be at the expense of international injustice. The exploitation of resources so as to improve the lives of poor present peoples may be at the expense of other species, and of future generations.

Like the local scale, these others all involve unequal power relations. But, whereas the people of poor localities and nations can raise some kind of voice, weak and muted though it may be, nonhuman beings and future generations are made all the more vulnerable by their silence. They cannot talk back, to engage the moral discourse creating environmental ethics.

SUSTAINABLE DEVELOPMENT

The notion of sustainable development is one of the most influential outcomes of interaction between environmental ethics and development ethics. It has important implications for social justice, especially with respect to future generations. The meaning of sustainable development is open to different interpretations (see Langhelle 1999). However, most of what is at stake is captured by the proposition that, although in the abstract it is concerned with sustaining the social, political and economic base of development, its core idea is that of a kind of development 'which so treats the natural environment that the process of development, or at least the products or benefits of that process, can continue into the future in a sustainable way, both for ourselves and our children, and for future generations' (Dower 1992: 93). The significance of sustainable development to international thinking was marked by the publication of the Brundtland Report in 1987, and consolidated at the Rio de Janeiro 'Earth Summit' in 1992.

Peet and Watts (1996: 1) point out that the notion of sustainability links three hitherto relatively disconnected discourses of global scope: those of environmental crisis, demography (the Malthusian spectre of population growth) and economic inequality. Recognition of their mutual interdependence is generating new perspectives on the relationship between development and environment. Among these is political ecology (Low and Gleeson 1998), which seeks to integrate land use practice in its broadest sense with political economy as the understanding of social practice at its most general level. Thus, interpretation of environmental crisis as a result of population pressure and resource mismanagement (especially by poor people) is yielding to an understanding of the role of increasingly global economic relations under capitalism in assisting well-to-do populations of rich countries to gain access to disproportionate shares of the world's natural resources. National politics and multinational organisations, like the International Monetary Fund and World Bank, are seen as implicated in this process, operating in the interests of increasingly advantaged persons and places rather than as neutral custodians of the general good.

These new understandings of the interrelated origins of underdevelopment and environmental problems have contributed to the rise of antidevelopmentalism and similar movements (mentioned in the previous chapter). These challenge previously dominant Western discourses from the perspective of postcolonial peoples – those of the poor 'South' rather than the privileged 'North'. The deeply normative character of these movements is underlined by the banners of liberation ecology, as 'a discourse about nature, Marxist in origin, poststructural in recent influence, politically transformative in intent' (Peet and Watts 1996: 37). Meanwhile, capital is by no means unaware of the threat to its conditions of production and reproduction, in the form of an environmental crisis, and to its domination of development discourse, in the form of opposition movements. The response involves more convincing gestures towards conservation, in capital's 'ecological phase' (Escobar 1996: 47).

The notion of sustainable development arose in reaction to these and other features of the contemporary context within which environmental ethics is theorised and practised. These include the scientific capacity to predict environmental impacts, within an increasingly sophisticated understanding of ecological systems and processes aided by such technical innovations as mathematical modelling and geographical information systems. The origins of this work go back to the reaction of geographers, regional scientists and planners to the requirements for environmental impact assessments in the United States under provisions of the National Environment Policy Act of 1969.

How such knowledge bears on moral responsibility and ethical understanding may be explained by a simple example (Gower 1995). Assume two isolated settlements on the banks of a river, the Greens upstream of the Blues and with industries polluting the river. As long as the Greens are unaware of the Blues and the effects of pollution there are no demands of justice, but knowledge creates awareness of obligation. The same holds for effects of one generation on another, separated by time instead of space. There can be differences in values which threaten an objective basis for moral responsibility, but obligations are nevertheless likely to be recognised:

> There may be many differences between the Greens and the Blues in what they believe about themselves, about their fellow citizens, and about the political, social and cultural structures they have created, but however extensive these differences may be, there will still be similarities. People in both communities will share a biological need for food, for clean air and water, for an uncontaminated environment, etc. Similarly in the case of non-overlapping generations. However great the differences in tradition and belief between us and future generations, some value will be placed on human life and upon the means necessary to sustain it.
>
> (Gower 1995: 55)

Something of the flavour of sustainable development as a normative project is provided by a strongly supportive text (Engel and Engel 1990). This notes the failure of major moral traditions to provide adequate environmental ethics for modern civilisation, and invokes 'the twin moral principles of social justice and environmental responsibility' (Engel 1990: 2). The first task is to understand different moral codes woven into cultures, leading to attempts to define a new social paradigm to promote sustainable development in each culture and region of the world. Sustainable is taken to mean not only indefinitely prolonged but also nourishing, so sustainable development is 'the kind of human activity that nourishes and perpetuates the historical fulfilment of the whole community of life on Earth' (Engel 1990: 10–11).

The opposition character of the project is stressed as follows: 'sustainable development is an authentic moral concern to the degree that it poses an alternative to the dominant model of modern development' Kothari 1990: 34). Four primary criteria are advanced for sustainable development as an ethical ideal: a holistic view of development; equity based on the autonomy and self-reliance of diverse entities instead of on a structure of dependence founded on aid and transfer of technology; an emphasis on participation; and the importance of local conditions and diversity. There are two broader considerations: a fundamentally normative perspective on the future, particularly from the viewpoint of the coming generations; and a cosmic view of life as sacred. Responsibility for the situation to which sustainable development is a reaction is firmly fixed:

> We have more than enough empirical evidence that the destruction of the biosphere lies first and foremost in the wasteful lifestyles of the world's privileged groups, and that the problem of poverty emanates from this same source. Consumption, as an end in itself, excludes the rights of others, both because it makes heavy demands on resources, but also because, in self-gratification, it is blind to others' needs.
>
> (Kothari 1990: 34)

An alternative philosophy of development is therefore needed.

> This calls for a perspective on science that is oriental rather than occidental, feminist rather than *macho*, rural rather than urban, one that draws on the accumulated wisdom of centuries . . . rather than one that rejects all that is past and traditional.
>
> (Kothari 1990: 35)

Calling on oriental thought is prompted by recognition that reverence for the natural world as a god's creation is incorporated in various non-Western religious traditions, along with the prevention of pollution as a functional necessity. Something of this is exemplified by Hinduism:

> Hindu culture, in ancient and medieval times, provided a system of moral guidelines towards environmental preservation and conservation. Environmental ethics, as

> propounded by ancient Hindu scriptures and seers, was practised not only by common persons, but even by rulers and kings. They observed these fundamentals sometimes as religious duties, often as rules of administration or obligation for law and order, but either way these principles were properly knitted with the Hindu way of life. In Hindu culture, a human being is authorized to use natural resources, but has no divine power of control and dominion over nature and its elements. Hence, from the perspective of Hindu culture, abuse and exploitation of nature for selfish gain is unjust and sacrilegious. (Dwivedi 1990: 211)

This provides a reminder that environmental ethics is not an exclusively Western intellectual creation. Indeed, there is a view that the world's indigenous religious traditions largely resonate with Leopold's land ethic (Callicot 1994). The problem is how to draw selectively on such traditions, adopting their ethical insights detached from the cultural specificity of their overall ethico-religious structure.

Calling on feminist perspectives invokes ecologocal feminism, or 'ecofeminism' (for further references see Jacobs 1995: 88–90; Booth 1997: 259–60; Cuomo 1998: 38–40). Ecofeminism draws on an analogy between male domination of women and of nature. Ecofeminists argue, among other positions, that women have different, more nurturing ways of seeing and relating to the world around them, which can provide insights into the treatment of nature. They also stress, and idealise, womens' experience of their own bodies as a bridge to nature: 'Women feel themselves as a part of the eternal cycle of birth, growth, maturation, and death, which flows through them, not outside them' (Pietilä 1990: 236). The ethic of care associated with feminism offers an alternative to a (masculinist) attitude to the exploitation of nature; it is taken further in the next section of this chapter.

The emphasis on rural rather than urban experience is suggestive of bioregionalism, mentioned in the previous section (for further references see Jacobs 1995: 90–2; Booth 1997: 257–9). This invokes living within the constraints and possibilities imposed by the character of a territory, as a human and natural community sustainable over time. Those who have learned to live in the region are considered best able to decide on its development. While bioregionalism may well provide a model for small-scale, self-contained and decentralised rural living, it does not have much relevance to life at the contemporary metropolitan scale, which draws on far-flung resources for its sustenance.

The notion of sustainable development is thus open to critique in some of its own terms. In addition, it has to manoeuvre between its claims to universal panacea and the evident local specificity of what may actually work, dependent as this will be on both culture and nature. Sustainable development has also been subjected to a more fundamental critique, suggesting that it promotes the agenda of multinational capital and the world's well-to-do (the North rather than the South), who have cleverly coopted much of the environmental movement, and that it places undue reliance on the market as arbitrator of conflict over the utilisation of resources. Hence the claim that what is at stake is not the sustainability of physical resources or local cultures but that of the global ecosystem 'defined according to a perception of the world shared by those who rule it' (Escobar 1996: 51–2). Growth and the environment have somehow to be reconciled, but it is growth as capitalist market expansion rather than environment which is to be sustained. In a similar vein is the suggestion that the international agenda for sustainable development is concerned with incorporating environmental assets into the economic system to ensure its sustainability, constructing environmental problems merely as efficiency issues to be managed more effectively (Doyle 1998: 774–5).

Socialism provides an alternative perspective for those who understand the contradiction between capitalism and environmental conservation (Pepper 1993). Ecosocialism stresses the mutuality of human and ecological well-being, but recognises a hierarchy of moral significance (Low and Gleeson 1998: 142); biocentrism is viewed as merely promoting certain human values, for example biodiversity and the protection of wilderness, at the expense of others such as the alleviation of poverty and social injustice. The concept of social reproduction, which features in Marxist theory and socialist practice, carries implications of sustainable development, for the reproduction of both labour and the means of production depend on access to a continuing supply of natural resources. That this might most effectively be achieved by their collective ownership or control is a position central to socialism, and the truth of which others may come to understand as free-market liberalism deepens the environmental crisis.

The pros and cons of sustainable development will not be followed further here. The purpose of introducing them briefly has been to underline the contextuality of this particular perspective, with its high moral tone. With so much at stake, for the world's rich and poor, it is not surprising that ethics becomes intertwined with (and perhaps clouded by) economics and politics. Sustainable development is, after all, a matter of social justice. What remains is to take up two issues already signalled above, those of community and care, and see what special significance they may have in looking after the future.

COMMUNITY, CARE AND THE FUTURE

> The vision of human life we ought to aspire to is not that of maximum exploitation of Earth as a big property resource, but that of a valued residence in a created community of life. (Rolston 1993: 278)

> For a society to be judged as a morally admirable society, it must, among other things, adequately provide for care of its members and its territory.
>
> (Tronto 1993: 126)

Of all the issues raised in this chapter, obligation to future generations is possibly the most intractable. Among the uncertainties are how far ahead to look and how long humankind will enable the future to be. The purpose here is to take up two notions that were given considerable attention earlier in this book, but which have been no more than hinted at in the present chapter, and to try to link them in a perspective on the future. These are community and care. Coupling them might provide the basis for an ethics of sustainability, which, while inevitably anthropocentric, is by no means inattentive to some of the intrinsic values attributed to nature in a biocentric perspective.

First to community. Leopold's land ethic, introduced earlier in this chapter, invoked an extended concept of community. This has been translated by some environmentalists (including deep ecologists) into a form of biocentrism in which individual entities have a purely instrumental role in system maintenance. Some of the objections to this have already been noted. But there is another reading of Leopold's understanding of community, as not strictly nonanthropocentric. He argued that the measure of land value is not merely the economic good of its owner; his notion of land as a community recognised complex connections among the elements, such that changes on the part of landowners could cause unanticipated damage for themselves and for others elsewhere. Thus:

> The communal meaning of land provides for a human communion with intrinsic features of nature. But at bottom, the human community with the land is based on the land's transmission, through natural processes, of the effects of human action on other human beings. Leopold's accomplishment is to bring consideration of the land into the ethical conception of the human community. (Trachtenberg 1997: 78)

The land is 'the substrate of human community', in which familiar ethical values such as not harming other people come into play. Natural processes are seen as part of the human connectedness recognised in communitarianism.

This sense of community is similar to that signalled by Rolston (1993: 278, above), who goes on to suggest that the Earth is valuable in a humanistic sense as a resource for people who are able to value it instrumentally, and in an ecological sense as able to produce value. This kind of understanding also seems consistent with the role of nature in the premodern concept of community in Africa, for example (see previous chapter), with an African world-view 'grounded in a conception of the self as intrinsically connected with, and part of, both the community and nature . . . the need to dominate nature as an impersonal object is replaced by the need to cooperate in nature's own projects' (Harding 1987: 360). Failure to do so could risk human survival, as such societies discovered in the ongoing practice of sustainable living.

A further insight from African land ethics introduces another dimension of community. The distinction between individual possession and communal ownership may be outlined as follows:

> The important thing which united all African societies with regard to ownership of land was that land was considered a communal property belonging to both the living and the dead. Those ancestors who had lived on the land belonged to the same social unit which owned and controlled the land, and each individual who used the land felt a communal obligation for its care and administration before passing it to the next generation. (Omari 1990: 168)

Such attitudes are by no means confined to African tradition. There is a tale from the Talmud, in which a rabbi saw an old man planting a tree and suggested that he would not live long enough to eat its fruit, to which the man replied, 'I found a fruitful world because my forefathers planted for me. So will I do the same for my children' (Palmer 1990: 58). In these and other traditions, reverence towards nature was both a religious attitude and a social practice, grounded in worldly prudence and material necessity.

The role of community in an intergenerational context has been elaborated by Avner de-Shalit (1995). He begins by proposing that we have both positive and negative obligations to close and immediate future generations, so we should supply them with certain goods as well as not burdening them with problems arising from our use of the environment. For very remote future generations our positive obligations fade away, though we retain a strong negative obligation to avoid causing them serious harm – comparable with those to geographically remote present people with whom we do not share community (de-Shalit 1995: 17). A theory supporting these intuitions will have to strike a balance between obligations to future generations and to our contemporaries, in particular the needy. As Langhelle (1999: 141) asks, 'Why bother about intergenerational equity if your own children have very poor chances of reaching adulthood?'

Communitarianism provides the basis of the theory de-Shalit advances. Recognising that the essence of community is continuity and succession, he proposes a 'transgenerational community' in which to ground obligation to the future. Cultural interaction and moral similarity among people over time create obligation to future generations, as part of our own constitutive community. This communitarian theory of intergenerational justice locates the source of our obligation to future generations in ourselves, who must take full responsibility for the state of the environment and for how it is left to future community members.

De-Shalit (1995) goes on to explain the defects of alternative moral perspectives which might be brought to bear on intergenerational justice. Conventional utilitarian formulations run into problems with the future population; examples include the fact that average utility or happiness can be increased by not giving birth to children likely to fall below the average. Problems with contractarianism arise from the impossibility of present and future people bargaining with each other or reaching a social contract. Grounding obligation to future generations in rights raises complications which do not arise with rights for present people: welfare rights can be envisaged, with implication for the present use of resources, but not such liberty rights as the freedom of speech of future generations which present people can hardly infringe.

The relationship between community and sustainability is a subject of continuing interest (Warburton 1998). It remains here to make the link with the notion of care. The religious traditions of premodern Africa and of Judaism, referred to above, both exemplify concern for the future expressed as an ethic of care for the land in a general sense. But the main source from which care has been imported into environmental ethics is feminist moral philosophy, central to which has been the ethic of care (see Chapter 5). The appropriation of an ethic of care in ecofeminism has taken a variety of forms. Examples are a 'partnership ethic of earthcare' involving equity between and moral consideration for human and nonhuman communities (Merchant 1996: 217), respect for cultural diversity and biodiversity, and inclusion of women, minorities and nonhuman nature in the code of ethical accountability (Low and Gleeson 1998: 148). Whatmore (1997: 45) identifies a 'politicised ecological care ethic', particularly among feminist environmentalist writers from a postcolonial perspective, which translates the recognition of webs of connectivity between the particular situated human actors and other biotic agents into acknowledged responsibilities, in the sense of both caring about generalised others and for concrete others.

Chris Cuomo (1998: 126) refers to the widespread acceptance among ecofeminists of a care ethic. But she is worried about talk of caring and compassion in the abstract, devoid of attention to the object of caring and the context in which it occurs. She shares with some others the fear of essentialism implicit in associating care exclusively with the female experience. Hence her qualified endorsement:

> An ecological feminist ethic that holds caring for other beings as good, and clearly sets out the appropriate objects and contexts of ethical caring relationships could certainly be informative and useful . . . men need to learn some ethical lessons from

> women's experiences as caretakers . . . given the historical, social, and cultural contexts, men ought to develop more caring attitudes, especially towards women and nonhuman entities. (Cuomo 1998: 130, 131)

A familiar theme is that men must stop trying to control nature and join women in identifying with nature.

An extended ethic of care exercised within a cohesive community could be a powerful force for both human and nonhuman betterment. However, any attempt to incorporate and fuse these two notions into a broader environmental ethic faces difficulties, recognised in the conventional versions of the perspectives of care and communitarianism (explained in earlier chapters), some of which are magnified in this new context. Care must be more than an optional exercise of empathy and charity (Low and Gleeson 1998: 22, 45). For its scope to be greater than parochial, it requires connection with an ethic of justice capable of responding to the uneven outcomes of interconnectedness on a global scale, and especially to the basic needs of poor present peoples. A communitarianism appropriate to the task must not be deluded by idealised notions of effortless harmony. Some associations between community (or nation) and nature (or land) can have exclusive or fascist tendencies, as in Nazi Germany (Harvey 1996: 170–1). And the notion of corporeal immersion of humans in the biosphere as the ultimate human/nonhuman community integration, while dissolving the infamous dualism of humankind and nature, poses enormous problems for human identity (Whatmore 1997).

There is also the perennial geographical problem of scale (Harvey 1996: 202–4). Should an environmentally caring communitarianism operate in a decentralised, local manner, along the lines of bioregionalism, or should it (could it) be global? If it is to work in practice it must be contextual, attending to local and particular details of complex processes. But external and spatially extensive linkages must also be recognised. It is in addressing this kind of question that geographical analysis (human and physical) and environmental science must consort with ethics if actual problems are to be addressed and possibly solved.

Meanwhile, the environmental crisis will not wait for ethics and science to catch up. At the present state of knowledge, there is much to commend theoretical pluralism and pragmatism, rather than commitment to some metatheory. King (1997: 226, 27) comments: 'Biocentrism as well as anthropocentric theories of human rights, justice, and care may function as tools in the assessment of moral situations, alongside other character traits and cultural norms particular to the situation', but warns that there are 'no built-in grounds for optimism that such a contextualist ethics will achieve consensus on ways of promoting human well-being that respect sustainable relationships with nature'. The best that can be expected is incremental enlargement of understanding, seeking a coherent reflective equilibrium (Stein and Harper 1996: 85).

But whatever the future of environmental ethics, and of life on this planet, recognition of caring within community as a virtue capable of incorporating nonhuman nature should be a moral imperative. As Harvey (1996: 197) reminds us, a virtuous relation to nature is closely tied to communitarian ideals of civic virtues. O'Neill (1996: 203–4) suggests that the distinction between environmental justice (as refraining from harming environments which provide the material basis for social life) and care for the environment goes beyond justice: it involves 'green virtues'. Given the antiquity of the question of how humankind should live with/in nature, it is not surprising to find echoes of the ethics of Aristotle's virtues and notion of flourishing deployed in this context, as if to complete a historical cycle of discourse. One version of ecological feminism drawing selectively (but not uncritically) on Aristotle provides a fitting finale to this chapter:

> Humans cannot flourish without other humans, ecosystems, and species, and nothing in a biotic community can flourish on its own. Likewise, communities (both social and ecological) depend on the existence of other communities. Ethical objects therefore flourish as both social and ecological entities. To be extracted from community, human or otherwise, is to lack relationships and contexts that provide the meaning, substance and material for various sorts of lives. (Cuomo 1998: 74)

CHAPTER 10

CONCLUSION: TOWARDS GEOGRAPHICALLY SENSITIVE ETHICS

> The wellspring of morality is the human capacity to put oneself in the place of others. (Paul, Miller and Paul 1994: vii)

Convention expects a conclusion. A book like this requires rounding off, but it can hardly be conclusive. It represents a brief passage in the ongoing discourse of creative scholarship, seeking to advance geographical understanding through the particular strategy of an engagement with ethics. Specific suggestions as to the place of ethics in geography and of geography in ethics may be found elsewhere (Proctor 1998a). This book, like its immediate predecessor (Proctor and Smith 1999), is more of an exemplar of the exploration of this new disciplinary interface, but with indications of the coherence which a more unified treatment might provide. It is offered in the hope that others might find this persuasive enough to continue the journey, though never to conclude it.

This final chapter recapitulates the significance of a world of difference to ethics, reinforced by reminders of some of the material presented earlier. It indicates the kind of moral knowledge that might be devised. It makes some suggestions about how a better world might be created. It finishes with observations on moral motivation, without which claims to the right and good will remain no more than utopian rhetoric.

A WORLD OF DIFFERENCE

Some of the most important debates of the final decade of the twentieth century, from social theory to cultural studies, swirled around the significance of difference. A politics of difference was deployed by various groups, mobilised under such banners as nationality, ethnicity, race, gender and sexuality, whose voices had hitherto been muted by what was held to be an essentialising and homogenising modernism. Conspicuously different identities began to flourish. However, there were accusations that the difference perspective placed too much emphasis on the symbolic process of culture, as opposed to the material process of political economy: domination appeared to be supplanting exploitation as the fundamental problem, with cultural recognition displacing socioeconomic redistribution as the remedy for injustice and the goal of political struggle (Fraser 1995: 68). Realisation that the age may be postsocialist as well as postmodern has tended to strengthen the sway of identity over class, along with reminders that the cultural and the material are intertwined (I. M. Young 1997).

Disquiet about difference nevertheless rumbles on. This is supported by such propositions that paralysis, crisis or Balkanisation rather than justice or democracy may be the consequences of the idea of difference, so long as external criteria by which to assess its claims are excluded, for ultimately there are as many differences as selves (Sypnowich 1993b: 105, 108). Undue reverence for difference encourages fragmentation of the world; along with depersonalisation of our relationships with others, this may be increasing a potential for evil 'not so different from that of earlier centuries' (Todorov 1999: 290). Ample evidence is available in the Balkans themselves, never mind Africa. Hence the reassertion of the significance of human sameness or close similarity, while accepting that some forms of difference are crucial to the way persons should be treated.

It is within this intellectual environment that geography is engaging with ethics. It is a situation within which the geographer should feel comfortable, given the discipline's long-standing attention to difference, and to the tension between an ideographic approach concerned with the unique and particular and a nomothetic approach concerned with the general and universal. It so happens that the spirit of the new times, in the aftermath of the nomothetic excesses of geography as spatial science, has been receptive to a recovery of the traditional preoccupation with the differentiation of humankind and its activities. In short, there has been a rediscovery of the significance of context in geographical knowledge.

The role of context to understanding morality has been central to the argument of this book. This is not a revelation confined to geography, of course, for it is in these recent years that ethics itself has been in part prised away from the dedication to universalism associated with its Enlightenment heritage, tentatively and selectively to embrace contextualism. In practice, however, moral philosophy seems more in tune with the historical than the geographical context, and disinclined to recognise the significance of such categories as location, place and space at anything other than a very general level.

Hence the significance of some of the work explained in earlier chapters of this book. Research conducted under the rubric of moral geographies (Chapter 3) provides thick (interpretative) descriptions of how moral values and practices vary from place to place, among neighbourhoods of cities for example. Facility locations, land uses and landscapes are capable of instructive moral readings, as are particular details of the built environment like memorials. These can be sites of conflict, as different moralities compete, and as the dominant discourse of a regime is challenged and overcome. Eastern Europe and South Africa provide plentiful evidence, only the tip of the iceberg of which was revealed in the case studies presented here.

Some of the most awful manifestations of conflict in the contemporary world arise from the (re)assertion of difference. Obvious examples are the ethnic and national chauvinisms of the Balkans, subdued under fifty years of communism but now seemingly determined to rewrite with renewed horror this region's claim to definitional status with respect to political fragmentation. Similar conflicts but with different resolutions are being played out in Israel/Palestine, as Arabs and Jews compete to establish 'facts on the ground' in order to influence an impending compromise of irreconcilable claims to territory. The struggle between ultra-orthodox and secular Jews for control of Jerusalem is but the latest twist in this region's ongoing saga of difference driving spatial separation (Chapter 6). Each of these conflicts has its own particularity, its historical and geographical context, understanding of which is crucial to any judgement of right or wrong, and any plans for peace which might pass the test of fairness.

Contemporary preoccupation with group identity and difference makes the resolution of such conflicts all the more difficult. Hence, again, moves to reinstate the significance of human sameness, stressed at various points in earlier chapters. Only when Arabs and Jews can understand their similarities and similar interests can a stable resolution be found for that part of the world. Separation may be a necessary expedient while hatred simmers. But as commentators on ethnoterritorial self-determination conclude: 'Most groups will need to find ways to live together in the same territory, rather than seek illusory territorial "solutions" to their ethnic conflicts' (Kymlicka and Shapiro 1997: 17). This is surely preferable to the 'protected spaces of many different sorts matched to the needs of the different tribes' (Walzer 1994: 78), which could describe what was euphemistically referred to as separate development under apartheid in South Africa. But it will require some innovative territorial-political structures and constitutions. Above all, spatial solutions will need moral imagination.

It is common to convey moral development as the capacity to overcome narrow parochialism and to adopt ever wider perspectives, and ultimately universal values. This is the point made by Robert Sack (1997) in his *Homo Geographicus*, and David Harvey (1996) in his urge to transcend local particularism. Similar arguments inform discussion of moral motivation, to be addressed at the end of this chapter. The tension between sympathy for close and familiar persons and universal beneficence (examined Chapters 4 and 5) took on special significance in consideration of the fate of Jews during the Holocaust. Glimpses into the debate on Jewish-Polish relations provided a demonstration of the role of context in the moral extremes of genocide and rescue, implicating both the local geography and the wider culture, but leaving room for individual human agency which defies spatial or structural determination. If there is any conclusion to be drawn, and entered into the debate on difference, it is this:

> [W]hat the Polish-Jewish experience suggests – to put it in the most general terms – is that 'identity politics' may be inadequate without a sense of solidarity. If we are to live together in multicultural societies, then in addition to cultivating differences, we need a sense of a shared world. This does not preclude the possibility of preserving and even nurturing strong cultural, spiritual, and ethnic identities in the private realm, not does it suggest collapsing such identities into a universal 'human nature'. But if multicultural societies are to remain societies – rather than collections of fragmented, embattled enclaves – then we need a public arena in which we can speak not only from and for our particular interests, but as members of a society, from the vantage point of the common good. (Hoffman 1998: 256–7)

That 'geography matters' is a contemporary catch-phrase of a discipline mindful of competing claims to relevance, and to resources in these mean times. To this local case of the significance of geographical context to human conflict and collaboration may be added broader examples. The political geography of nation states, formed so recently in the long span of human history, is already proving inadequate to cope with the globalising scale of economic and environmental processes. Among the specific problems are: threats to the spatial framework within which rights and obligations have usually been implemented, in the context of national citizenship (Low and Gleeson 1999); difficulties defining political jurisdictions fair to individuals, communities, minorities and parties within nation states, and in international organisations like the European Union (Johnston 1999); and recognition that a system of national sovereignties 'is not well adapted for harmonious relations with the Earth commons' (Rolston 1993: 267).

Postmodern perspectives make much of the notions of fluidity and hybridity, with the mobility of persons and the complexity of their identities seen as constrained by fixed boundaries or categories. The issue of open or closed borders is an important geographical case, introduced in earlier chapters in the context of communitarianism and control of space. Poland again provides an example of differences in mobility with implications for other postsocialist countries and indeed for the globalising world as a whole. Western capital is discreetly buying Polish farmland, much cheaper than in adjoining Germany, anticipating Poland's entry into the European Union in which pecunious citizens of all member states are free to purchase land, houses or anything else in other member states. The prospect of Germans buying farms in Poland raises the spectre of past invasions, Polish blood spilled fighting for their land and so on, reminiscent of the controversy over a Belgian Bank dominating sacred space in central Warsaw (see Chapter 3). Hence the debate over whether foreign purchase of land in Poland should be restricted (Ascherson 1999). After all, there are severe limitations on Polish workers moving to Western Europe to compete for jobs. Why should capital, which Marx described as 'dead labour' (in recognition of the labour power which created it), have greater freedom of movement than living labour?

Of course, there is a technical answer: that capital in the form of money can be moved from place to place with the minimal cost of an electronic computer transaction. Workers are less mobile, even in the multinational setting of the European Union, because they are not inanimate factors of production but persons, culturally embedded and often with deep attachment to place. This is one of the many reasons why the perfect markets idealised in economic theory cannot work in practice. And as capital (or its owners) are increasingly free to realise the benefits of its near-perfect mobility, the good fortune or otherwise of the workers is dependent on how their place is evaluated by market forces beyond their control.

Not all persons are immobile, of course. One of the revolutions of modernity was potentially to release people from the confines of limited opportunity and of local moral conventions. Some predictions of winners and losers in the coming world order have been made by the foundational head of the European Bank Reconstruction and Development, Jacques Attali, who envisages a future of 'rich nomads' and 'poor nomads' (Doyle 1998: 784). The rich, as the consumer citizens of the world's privileged regions, will be able to participate in the liberal market culture of political and economic choice, roaming the planet in search of the information, sensations and goods that only they can afford, while yearning for lost human fellowship, home and community. They will be confronted by the poor, roving masses of boat people on a planetary scale, seeking escape from the destitute periphery where most of the world's population will continue to live. Their encounters will be a far cry from the rich cultural diversity of the porous places sometimes celebrated in postmodern accounts word-processed in privileged palaces of academic life.

Imaginative as Atali's scenario is, the reality is likely to be a more spatially fixed polarisation, with extremes of rich and poor juxtaposed at different geographical scales. The local manifestations are suggested no more clearly than in post-apartheid South Africa (Chapter 8). Here the well-to-do (mostly 'white') occupy increasingly separate well-fortified homes and residential enclaves, armed instant-response services available to tackle intruders, the inhabitants emerging apprehensively for increasingly hazardous journeys to work, school and shopping centre with their risk of car hijacking and worse. The poor (almost exclusively 'black'), swamping the cities with fading prospects of sustainable living from the soil, find whatever spare space there may be on which to build a shack as the only alternative to the squalid township slums, casting around for means of subsistence increasingly confined to the informal economy, scavenging and crime. As formal law enforcement fails to cope, local vigilantes take their place. Nationally, the state's forces of social control maintain some semblance of order, until the dam bursts.

It is to deny such futures, and to promote more humane visions, that morality matters:

> The faith that the future is open and contingent in important respects upon human choice, that individual persons can identify with people in other cultures, with the dispossessed, with the community of life on Earth, and that humans have the capacity to make universal moral judgements, is essential to any possibility of creating a new world order. (Engel 1990: 9)

That this faith is being challenged by both the resurgent absolutisms and the widespread relativism of our time points to the tough challenge ahead, if a geographically sensitive ethics is to help create a better world.

CONTEXT-SENSITIVE MORAL KNOWLEDGE

Ethical argument sometimes proceeds as if there was a choice to be made between embracing universal principles available for application to specific situations, and accepting that the local particularity of moral beliefs makes implausible any transcendent values to which they can be referred. Fortunately, some distinguished moral philosophers, and an increasing number of geographers, understand that this is one of the false dualisms with which social thinking appears to be plagued. Otherwise, it would be hard to see how anything resembling moral truth might be discovered, or created, except universals so vacuous as to be impotent in the face of our sorry reality, or beliefs so restricted in scope as to be part of the problem rather than part of the solution of human conflict.

What is less clear is how to seek something capable of doing the work of moral truth, in a world of difference where context sensitivity may be the most obvious universal proposition. We could begin with the given morality of the individual's community of fate, into which one is born, but which must not be totally confining otherwise it would simply be reproduced. Thus:

> [T]he fact that the self has to find its moral identity in and through its membership in communities such as those of the family, the neighborhood, the city and the tribe does not entail that the self has to accept the moral *limitations* of the particularity of those forms of community. Without those moral particularities to begin from there would never be anywhere to begin; but it is in moving forward from such particularity that the search for the good, for the universal, consists.
>
> (MacIntyre 1985: 221)

If 'there are the makings of a thin and universalist morality inside every thick and particularist morality' (Walzer 1994: xi), this provides possible means of communicating with others elsewhere, in terms of those grand moral concepts like justice and liberty which will be universally recognised, even if what they mean in practice depends on the local and particular context. Culture-specific ideas of particular ethical and religious traditions may be turned increasingly universal:

> Ideas of how things 'ought to be' have slowly evolved in various writers and traditions. Such theories have developed and broadened and deepened, and collective human action has tended to make them clearer, more insistent, more definite and universal. Our notions of human dignity today are fuller and more articulated than the earliest ones, with which they remain continuous. The historical process itself has redefined and refined the meaning of being human, all the while expanding it, and has created the material and political conditions for actualizing this in human social relations. (Aronson 1990: 91)

The case study of South Africa (Chapter 8) provides an example of what is involved in working between the universal and the particular. It was explained that development as change for the better in the post-apartheid situation might initially be defined in terms of racial redistribution, with the imperative of satisfying basic needs of the poor, black population. This is a context-specific version of a universal principle of social justice as a process of equalisation, in the interests of the worst-off. However, bringing this to bear in the particular circumstances of contemporary South Africa revealed limitations of a narrowly (re)distributive perspective. This led to reflections on an alternative perspective, which replaces individual material well-being by care for other members of an emergent national community as the moral priority. Further consideration of the context revealed the difficulty of advancing such a conception in practice. All this is part of an ongoing dialogue between what might claim to be a universal moral perspective and the context within which it has to work, if it is to improve real impoverished lives.

Another example is provided in the case study of the moral landscape of the Polish industrial city of Łódź (Chapter 3). An important feature of capitalism is the control which an enterprise will seek to exert over its labour, in the interests of efficiency and profits. That this may take distinctive spatial forms is illustrated by the factory township, brought to heights of paternalist sophistication in the textile mill complexes of nineteenth-century Łódź. Evidence of how these particular complexes were operated contributes to knowledge of the general process of exploitation under capitalism. Added to this is the insight that the socialist enterprise sought similar control over urban space, populations and political processes relevant to its own success (Domański 1997). A spatial strategy originally considered distinctive to capitalism might then be interpreted as universal to modern industrialism, changing its moral interpretation.

Converging on something with claims to moral truth involves generalisation with increasing (spatial) scope. The approach is essentially pragmatic. It is the same as was adopted in exploring geographical aspects of social justice (D. M. Smith 1994a: 16), citing the following rationale:

> Many moral philosophers today see ethical theory as a matter of working back and forth between our moral intuitions about particular cases and the general principles that account for those particular judgements or are themselves intuitively attractive, in order to weave our moral thinking into a coherent and consistent web.
>
> (Arthur and Shaw 1991: 10)

General principles work with other factors in moral justification: 'We weigh one against the other, sometimes modifying principles and other times intuitions, in a search for a coherent reflective equilibrium' (Stein and Harper 1996: 85). The notion of reflective equilibrium was introduced by John Rawls (1971: 21), who argued that a conception of justice cannot be deduced from self-evident premises or principles; instead, 'its justification is a matter of the mutual support of many considerations, of everything fitting together into one coherent view'. In other words, 'principles of justice are mutually supported by reflecting on the intuitions we appeal to in everyday practices, and by reflecting on the nature of justice from an impartial perspective that is detached from our everyday position' (Kymlicka 1990: 70). Case material is involved in a process whereby intuition or considered convictions about real situations interact with more abstract deliberations.

Thus, in the present volume an egalitarian conception of social justice acted as a starting point for an approach to the reality of post-apartheid South Africa, modified in the process of this encounter, until the distributive conception of justice was itself found too narrow for the task at hand. This finding fed back into an understanding of social justice as requiring reference to the meaning of the good life, leading to the conception of the good suggested by African experience concerning care for those in greatest need within some form of community. This calls on understanding from premodern ethics as well as the contemporary ethics of care and of communitarianism. A perspective reflecting these positions was subsequently applied to environmental ethics and intergenerational justice (Chapter 9), with tentative findings which might be played back into the emerging general theory. And so it goes on.

The approach adopted is deliberately pluralistic and pragmatic, calling selectively on such moral theories as liberal egalitarianism, communitarianism and the ethics of care. This strategy seems consistent with tentative early stages in the exploration of a geographically sensitive ethics. The perspective of this book resonates closely with the account of fallibilistic moral pluralism provided by Lawrence Hinman (1994: 48), incorporating 'sensitivity to the contextuality of our moral beliefs and the recognition that moral disagreement and conflict are permanent features of the moral landscape', while believing that some moral positions are better than others. He recognises the problems of practical application: 'one of the single greatest difficulties that act-oriented moral philosophies face is in applying a moral theory to a particular case. A morally sensitive character is more likely to ensure that we apply a principle with insight and creativity' (Hinman 1994: 315). The case study is central to geographical method, and its utilisation is crucial to the development of geographically sensitive ethics. The reference to character underlines the insight, sometimes attributed to feminist moral philosophy, that feeling as well as reason is involved in making moral judgements and engaging ethics, as in other aspects of life.

Philosophers are increasingly recognising that, if moral theories are to be of any real help in an increasingly complex and changing world, they must become more empirically informed (Louden 1992: 127). Philosophical acumen is necessary for understanding values and value-laden phenomena. But 'philosophy alone, uninformed by social science, loses touch with empirical contingency and variation and with the insights to be gained from close study of actual experience' (Selznick 1992: xiii). And in this respect, there can be no more effective research practice than that of geography.

TOWARDS A BETTER WORLD

> The point of morality is not to mirror the world, but to change it.
>
> (Williams 1972: 47)

I have been reticent so far in the expression of my own moral values. Some have seeped into arguments in earlier chapters, while broader metaethical positions are reflected in the approach to moral knowledge recounted above. However, the purpose has been more to exemplify a geographically sensitive ethics than to provide moral prescriptions, far less anything resembling a coherent moral theory. Some lines of moral argument suggested in the body of the book may now be made more explicit.

The heart of the book is in the exploration of social justice (Chapter 7), and in particular in the argument concerning the place of good fortune. Luck plays an important part in people's lives, no more so than in the chance of birth in a particular place. With this alone come unequal opportunities, which are arbitrary from a moral point of view. Under the particular form of economic arrangements known as capitalism, these initial advantages, themselves inherited from earlier privilege, are easily passed on to future generations to compound the benefits, just as the burdens arising from the place of bad fortune may be perpetuated. Three interlocking competitive inequalities, concerned with natural resources, migration and bargaining power, are crucial contributors to international differences in individual life-prospects (Miller 1992: 304). They are similarly implicated in the more local patterns of inequality and polarisation which typify capitalist society. And, lest socialism as actually practised be thought of as benignly egalitarian, there is ample evidence of the transmission of advantage with a spatial expression in Eastern Europe, particularly within cities (D. M. Smith 1989; 1994a: Chapter 7).

Some observations concerning human sameness lead to recognition of common human needs, the satisfaction of which, for everyone everywhere, should have high moral priority. If the notion of human rights means anything, then it must surely cover the universal right to life manifest in basic need satisfaction. The claim that a universal conception of basic human need imposes some homogenising modernism, which denies human difference, is either ignorance or wilful diversion. As Stuart Corbridge (1993a: 469) explains, the 'postmodern dilemma' is avoided by accepting that certain human needs and rights are universal, and that in attending to them 'we are not so much dictating to others as dictating to ourselves'. What we are dictating is a moral imperative of self-sacrifice for the sake of the worst-off.

Another social condition which might claim to be a basic human need is security. This may be associated with a safe place in a geographical sense, and within a social structure. Security enables persons to plan and engage in long-term projects, as opposed to merely responding to pervasive unpredictability (MacIntyre 1985: 103–4). Collective action, in the form of community projects or national welfare programmes, is an obvious way of reducing insecurity. The atomisation or individualisation eroding community cohesion, along with the contemporary resurgence of neoliberalism threatening welfare states, undermine important sources of individual and group security. In particular, the free-market fundamentalism associated with the Reagan and Thatcher years and perpetuated in much of the political-philosophical baggage with which entry into the new millennium is encumbered, is a serious and growing source of insecurity. This is manifest in the closing of factories, loss of jobs and devastation of communities, at the hands of forces which in the globalising economy are increasingly beyond local influence and control.

This is not the place for a critique of free-market economics and neoliberal political philosophy, some of the main points of which have been made elsewhere (D. M. Smith 1994a: 279–84). The conspicuous outcomes in the form of inequality and insecurity may be indictment enough. All that need be added here is that initial inequality in market power can generate further inequalities, perpetuating and exacerbating patterns of uneven development for which there is no moral justification. The model of the ideal self-regulating, welfare-maximising market simply does not work as it is portrayed in textbooks, for reasons which include the impossibility of such features as perfect factor mobility and free competition in the real geographical world. Its perpetuation has much to do with the fact that the ideology of free markets tends to serve the interests of the well-off. As to environmental impacts, the imperative of producer survival under competition ensures that the market is preoccupied with the short run: 'the market ignores the needs of future people' (de-Shalit 1995: 135).

However, market forces are by no means the only threat to security. Contemporary ethnic and national chauvinism in the hands of those asserting the superiority of their difference suggests that the propensity for human evil is impervious to the civilised values which much of humankind proclaims, calling into question notions of moral progress. When the dust settles, there will be something to learn from the barbarity of the Balkans, as there has been from the Holocaust, not least about the significance of spatial relations like neighbourhood and community in the treatment of different others. But it will be too late to save hundreds of thousands of people unfortunate enough to have been who they were and in the wrong place at a crucial time.

Another aspect of the Balkans conflict with moral lessons is the conduct of warfare. In early 1999 NATO forces engaged in aerial bombardment of Serbia, 'delivering ordinance' (official warspeak for bombing) to 'degrade' (destroy and kill) armed capacity to undertake further ethnic cleansing in Kosovo. The preference for air raids rather than ground warfare minimised the risk of NATO casualties, but increasing 'collateral damage' (non-combatants killed or injured) gave new meaning to 'clinical bombing' and 'surgical strikes'. This strategy followed the contemporary practice of conducting technologically sophisticated warfare in a way that distances those responsible from its consequences, just as the lexicon of euphemisms facilitates verbal distancing. For example, a British pilot bombing Iraq in 1998 described himself as being in his 'own little electronic world', with only the occasional glimpse of flashing light outside (BBC Radio 4 News, 19 December 1998). However, modern means of telecommunications brought into American and British homes images of Serbian civilians killed by mistake, to add to those of the burning houses, mutilated corpses and refugee streams of Kosovo Albanians. It is within these different and competing spatialities of experience that moral judgements now have to be made, as to the conduct of warfare and its consequences.

The kind of concerns summarised here do not include all those exercising the moral indignation of some exponents of the politics of difference. This is not to suggest that contemporary preoccupations with constraints on freedom to indulge individual difference are of no consequence; it is that some of the injustices identified are endured in the relative material comfort of affluent societies, their victims otherwise protected from some of the deprivation perpetrated on multitudes of vulnerable persons in other places. Thomas Nagel (1993: 83) asks whether the subtle refinements that worry the inhabitants of liberal democracies, in which the most basic protections of the individual are taken for granted, even belong to the same subject matter as the concerns of those in countries where basic human rights are not upheld: 'Is there any meaningful sense in which freedom from torture and freedom to rent pornographic videos both raise an issue of human rights?' Such a question provides another reminder of the contextuality of moral values and priorities.

In summary, the moral message of this book is the imperative of developing more caring relations with others, especially those most vulnerable, whoever and wherever they are, within a more egalitarian and environmentally sustainable way of life in which some of the traditional strengths of community can be realised and spatially extended. This awesome task achieved, many lesser problems would doubtless fall into place.

ON MORAL MOTIVATION

> Unless truths of morality can be identified by reason, moral conflicts are only clashes between people with different unverifiable beliefs. Then, victory goes to the side with the power (of arms, numbers, persuasion, habit, or tradition) to prevail, and right becomes indistinguishable from might. (Reiman 1990: ix)

> Justice begins with and presupposes our emotional engagement with the world, not with philosophical detachment or with any merely hypothetical situation.
>
> (P. Singer 1995: 266)

None of the aspirations outlined in the previous section entail moral motivation, in the sense of inducing people to act in pursuit of a more caring and equal world. Why should we care, about those less fortunate than ourselves, viewed from our personal pinnacles on the highly uneven surface of human well-being? Why should we promote equalisation, when we ourselves along with our nearest and dearest are likely to be losers in the redistribution of resources? Why bother with what might appear to be little more than utopian idealism, in a world which seems impervious to change in conscious pursuit of the good?

These metaethical issues arise in any project which, from the perspective of the relatively advantaged parts of the world, raises the question of how to extend the spatial scope of beneficence. Neither external forces nor something internal to the self, neither the abstract reason of intellectual persuasion nor the immediate experience of empathy with close people, seem sufficient in themselves. If there is an answer, it must have something to do with understanding what suffering actually means, arising from some combination of personal experience and the capacity to generalise from the experience of others. How this perspective is gained may depend very much on where we are in relation to others, on how engagement with others is mediated by geographical space.

Similar pointers may be drawn from writers with otherwise different philosophical positions. For Richard Rorty (1989: 192), moral progress is in the direction of greater human solidarity: the ability to see traditional differences – of tribe, religion, race, customs, and the like – as unimportant when compared with similarities with respect to pain and humiliation, 'to think of people wildly different from ourselves as included in the range of "us"'. However, this is achieved not by inquiry but by the imaginative ability to see strange people as fellow sufferers: solidarity is 'made rather than found, produced in the course of history rather than recognized as an ahistorical fact' (Rorty 1989: 195).

Onora O'Neill (1996: 197) introduces a geographical dimension to solidarity, which 'can be expressed across large spatial and social distances, to others who are neither near or dear, through forms of help and support for distant strangers, especially those who are destitute or oppressed'. Examples include the former anti-apartheid movement, and charities providing aid to the Third World (Silk 1998). She goes on to point out that acts of rescue are more dramatic expressions of care and concern, directed towards others in present danger and misfortune. The discussion of rescue of Jews during the Holocaust indicates how difficult it is to generalise about the motivation for this most personal, and potentially hazardous, kind of beneficence, though the capacity to think in universal terms was involved, along with the ability to adopt the perspective of the other (see Chapter 5).

Peter Singer stresses that to live ethically is to act on reflection, which if done properly will lead to the conclusion that well-being requires commitment to a transcendent cause, to the point of view of the universe rather than sectional interests.

> From this perspective, we can see that our own sufferings and pleasures are very like the sufferings and pleasures of others, and that there is no reason to give less consideration to the sufferings of others, just because they are 'others'.
>
> (P. Singer 1995: 222)

Norman Geras focuses on the experience of both suffering and well-being in engendering a sense of solidarity:

> [B]asic codes of moral conduct and deliberations about what may or may not be rightly done relate in definite, if not always direct, ways to considerations of suffering – the avoidance and alleviation of it – and to the promotion and maintenance of well-being, in the largest sense. This is the type of consideration which nearly anyone can understand as the possible candidate for a compelling, action-guiding reason, because in the experience of everyone will be some kind of knowledge, albeit of variable breadth, depth and sensitivity, of what suffering and well-being actually feel like. (Geras 1995: 94)

Hence the importance of combining reason and feeling, intellectual abstractions and experienced empathy, indeed of recognising their inseparability in moral life (as suggested in Chapter 5).

Bryan Turner (1986: 97) posits: 'If there is a universal emotion it may well be a sense of outrage which emerges from our experience of injustice when the innocent are overwhelmed by superior forces', a transcendent experience which cannot be relativised. This is something which can be understood, because humans have the ability to put themselves in the place of others. Experiencing injustice, perhaps victims ourselves, turns into a sense of justice, 'as we learn to see ourselves in others' places and realize that we are never mere observers in injustice but almost always at least passive participants' (Solomon 1995: 278). In his *Theory of Moral Sentiments*, Adam Smith proposed that we have a natural tendency to place ourselves in the situation of others, to take an interest in their interests (Kukathas 1994: 12). Turner (1986: 115) goes on to suggest that 'although inequality constantly re-emerges in human societies we appear to have a "natural" sense of fairness and justice which develops out of the reciprocity that determines the contours of everyday life'. Thus opposition and resistance to inequality seem as inevitable as inequality itself, knowledge of which is sufficient to motivate action – at least for some people in some places.

Chris Cuomo stresses the importance of extending moral concern, from the self to immediate contacts and beyond:

> [E]motional, symbolic, and cultural connections with others help move us beyond simple egoism and generate concern for and motivation toward the interests of others . . . The challenge for ethics is to capitalize on the fruitfulness of given, obvious connections within our species and communities without representing these connections as the summation of moral life. (Cuomo 1998: 49–50)

A similar line can be found in the framework for understanding *Homo Geographicus* elaborated by Robert Sack (1997; see Chapter 1): as understanding of our interrelated world widens, awareness of the implications of our actions provides a basis for their evaluation and a reason for action, a source of moral motivation. Stuart Corbridge (1998: 37) similarly locates moral responsibility in an understanding of the way in which the unfortunate lives of distant strangers are bound up with our own, in the globalising world economy: 'there are good reasons for attending to their needs and rights as fellow human beings in a manner that will make calls upon "our" resources and entitlements'.

Such reasoning may be supported by Kant's categorical imperative, but as a thought experiment in role reversal rather than merely an exercise in abstract reason. Insofar as it can address real others, the categorical imperative 'applies to maxims that are replete with the local, partial, conditional, and contextual' (Sack 1997: 232). The view from behind Rawls's veil of ignorance as to people's status in society invites a similar interpretation:

> Rawls's theory of justice is most coherently interpreted as a moral structure founded on the equal concern of persons for each other as for themselves, a theory in which empathy with and care for others, as well as awareness of their differences, are crucial components. (Okin 1989: 248)

Others read Rawls in this way:

> Rawls forces us to put ourselves in the position of those people who, for no good reason, are less well off than ourselves . . . the needs and rights of strangers could easily – and but for the 'accident' of birth, be the needs and rights of ourselves.
> (Corbridge 1993a: 464; see also 1998: 45–6)

Hence the expression of such sentiments as 'there but for the grace of God go I', 'if the tables were turned', 'put yourself in my place', and so on, in everyday moral discourse.

Some find community an essential ingredient in moral motivation. Asking what it requires to be both other-regarding and self-preserving, Philip Selznick (1992: 183–4) responds: 'Both ideals are attenuated when virtue is perceived as abstract or rule-bound; they are strengthened insofar as organic ties to persons, history, deeds, and nature are created and sustained.' His quest is for community that looks outwards rather than inwards: 'A crucial step is the embrace of strangers. When strangers are treated with the respect due members of an enlarged community, a moral watershed is reached' (Selznick 1992: 196). He adds: 'Virtues of caring, fidelity, and reverence – for persons, not abstractions – may flourish in many settings' (Selznick 1992: 206). These settings all have their geography, their spatial form and relationships.

Few moral philosophers, even those of a communitarian persuasion, have anything to say about spatial relations. Selznick (1992: 193) notes: 'The sociological lesson is not that the clock must be turned back. It is, rather, that we must now more frequently acquire by sensitive design what once could be taken for granted.' Hence the attempts, with varying degrees of success, to (re)create the intimacy and mutuality of ideal communities in the 'urban village', the planned neighbourhood unit of new towns, the *mikroraion* (microregion) of Soviet cities, and some experiments with communal living like the Israeli *kibbutz*. These days such projects may encounter suspicion of the 'modernist' idea that physical structures can significantly influence human behaviour in the direction of collectivism and mutual responsibility. However, the idea that human motivation is impervious to environment is no more defensible than determinism.

Selznick (1992) and others are attracted by arguments for intricate and close-grained diversity to protect the vitality of spontaneous human relations in an urban setting. For example, the ideal of city life proposed by Iris Marion Young (see Chapter 5) involves 'the being together of strangers ... a side-by-side particularity neither reducible to identity nor completely other', with groups intermingling without becoming homogeneous, and people 'open to unassimilated otherness' (I. M. Young 1990b: 237–41). The problem with such a scenario is its departure from the reality which it would have to replace, before reaping the anticipated moral harvest of inter-group harmony. For a rigorous testing ground, she could try Jerusalem.

That face-to-face contact remains important to human interaction is a matter of everyday experience. What, then, are the implications of some contemporary social trends, such as growing numbers of one-person households and the proliferation of mobile 'phones? This leads to questions about how remote interaction might affect interpersonal behaviour. A research project conducted at the University of Pittsburgh into the use of the Internet found a decline in interaction with family and a reduction in the circle of friends, that directly corresponded with time spent online (*Guardian*, 31 August 1998). However, supportive relationships can arise from initially remote interaction (Rheingold 1993). Cyberspace has also enabled the creation of worldwide national identity communities among members of national diasporas, which may serve to strengthen and separate them at a global rather than local scale (Newman and Paasi 1998: 199).

So, what kind of 'communities' or groups may come to exist in cyberspace? What kind of morality is likely to emerge, between the extremes of universalism and partiality? Can the 'annihilation of space' dissolve indifference to the fate of far distant others, or will physical separation continue to be a barrier to care and compassion? Could remote communication promote evil, with virtuality substituted for reality, or reality responding to virtuality, or persons lost in confusion between them? And what of individual volition in new contexts, when people can still be described as 'born bad'?

In 1996, the people of Dunblane in Scotland were coming to terms with the mass murder of school children by a gunman. They were asking how this could happen, in a place some described as a 'supportive' and 'tightly-knit' community, but which to others was an anonymous commuter suburb. What kind of 'community' was it? What was really behind the action of a perceived 'oddball' or 'misfit', somehow not knitted-in, whose different otherness was finally resolved in the violence of mutual annihilation? One of a series of shootings in the United States in 1999 (at Columbine High School, Colorado) revealed a similar story of difference, of 'loners' turning their explosive rage on members of cliques who rejected them, and then on themselves. That the overwhelming response to such events focused on tighter gun control (sensible though this would be), rather than on what leads someone to kill, underlines the limited grasp of context on the part of those seeking to prevent repetition. While their full interpretation is beyond our concerns, such cases often arise from a failure of socialisation, someone not bonding with a loving other in early life and deprived of the experience of empathy, like some of those who commit war crimes. Returning to some notions from Chapter 4, they may reflect a person incapable of constitutive attachments, without moral depth (Sandel 1992: 179), or withdrawn from moral judgement, ceasing to interact in the human community (Benhabib 1992: 125–6).

So, where does this leave us? Not with some reliable geographical fix for evil, or for the promotion of beneficence, important though spatial relations may be. Societies are powerless to prevent the occasional moral aberration like mass murder. But their incidence might be reduced by a sustained assault on those forces which make for broken families and lives: poverty, insecurity and the hopelessness of not belonging (along with better mental health services, for those who still get hurt). Also helpful might be less of the assertion of individual and group difference and consequent exclusion, as part of the struggle for identity, more attention to the incorporation and participation of all in the sociability of human life, less energy dissipated in divisive competition and more devoted to cohesive collaboration.

Above all, some of the lost moral certainties of earlier ages need to be recovered, not as fixed and repressive codes, but as carefully crafted anchors to prevent continuing drift in a sea of relativism or nihilism:

> If there is no truth, there is no injustice. Morally and politically anything goes. There are appalling language games always in preparation somewhere, now as much as ever. They will be 'played' by those looking for the chance of it in deadly earnest. It remains to be shown that, amongst our defences against them, we have anything better than the concepts of a common humanity, of universal rights, and of reasoning together to try to discover how things are, in order to minimize avoidable suffering and injustice. (Geras 1995: 143)

The discovery of how things are, and how they might be improved, will be the more sensitive to those differences that really matter the more geography is involved. And if a geographically sensitive ethics has no more than one major message, it is, again, the importance of context, of understanding the particular situation: how things are, here and there. If the human capacity of putting one's self in the place of others is to be an effective wellspring of morality, this requires understanding that place, as well as those others.

BIBLIOGRAPHY

Adam, H. (1994), 'The Mandela personality cult', *Indicator South Africa*, 13(2), pp. 7–12.

Adams, P. C. (1996), 'Protest and the scale politics of telecommunications', *Political Geography*, 15, pp. 419–41.

Adelson, A. (ed.) (1996), *The Diary of Dawid Sierakowiak*, London: Bloomsbury.

Adelson, A. and Lapides, R. (eds) (1989), *Lódź Ghetto: Inside a Community under Siege*, Harmondsworth: Penguin.

Adelzadeh, A. (1996), 'From the RDP to GEAR: the gradual embracing of neo-liberalism in economic policy', *Transformation*, 31, pp. 66–95.

Adelzadeh, A. and Padayachee, V. (1994), 'The RDP White Paper: reconstruction and development vision?' *Transformation*, 25, pp. 1–18.

Almond, B. (1991), 'Rights', in P. Singer (1991), pp. 259–69.

Almond, B. (1995), 'Rights and justice in the environment debate', in Cooper and Palmer (1995), pp. 3–20.

ANC (1994), *The Reconstruction and Development Programme*, Johannesburg: African National Congress.

Anderson, E. S. (1999), 'What is the point of equality?' *Ethics*, 109, pp. 289–337.

Anderson, P. J. (1996), *The Global Politics of Power, Justice and Death: An Introduction to International Relations*, London: Routledge.

Andrusz, G. (1996), 'The "post-socialist" city: from state to market', *City*, 1/2, pp. 23–9.

Ardrey, R. (1971), *The Territorial Imperative*, New York: Dell.

Armstrong, K. (1996), *A History of Jerusalem: One City, Three Faiths*, London: HarperCollins.

Aronsen, R. (1990), *'Stay out of Politics': A Philosopher Views South Africa*, Chicago: University of Chicago Press.

Arthur, J. and Shaw, W. H. (eds) (1991), *Justice and Economic Distribution*, 2nd edition, Englewood Cliffs, NJ: Prentice Hall.

Ascherson, N. (1999), 'The ghosts of past invasions that still haunt Poland', *Guardian*, 21 February.

Atkinson, D. (1991), 'The state, the city and political morality', *Theoria*, 78, pp. 115–38.

Attfield, R. and Wilkins, B. (eds) (1992), *International Justice and the Third World*, London: Routledge.

Azaryahu, M. (1996), 'The spontaneous formation of memorial space. The case of Kikar Rabin, Tel Aviv', *Area*, 28, 501–13.

Badcock, B. (1998), 'Ethical quandaries and the urban domain', *Progress in Human Geography*, 22, pp. 586–94.

Bader, V. (1995), 'Radical democracy, community, and justice: or, what is wrong with communitarianism?' *Political Theory*, 23, pp. 211–46.

Baier, A. C. (1987), 'The need for more than justice', *Canadian Journal of Philosophy*, 13, pp. 41–56.

Baker, J. (1987), *Arguing for Equality*, London: Verso.

Bar-Gal, Y. (1985), 'The *shtetl* – the Jewish small town in eastern Europe', *Journal of Cultural Geography*, 5(2), pp. 17–29.

Bar-Gal, Y. (1991), *The Good and the Bad: A Hundred Years of Zonist Images in Geography Textbooks*, Research Paper 4, London: Department of Geography, Queen Mary and Westfield College, University of London.

Barry, B. (1989), *Theories of Justice: A Treatise on Social Justice, Vol. 1*, London: Harvester Wheatsheaf.

Barry, B. (1995), *Justice as Impartiality: A Treatise on Social Justice, Vol. 2*, Oxford: Clarendon Press.

Barry, B. and Goodin, R. E. (eds) (1992), *Free Movement: Ethical Issues in the Transnational Migration of People and Money*, New York: Harvester Wheatsheaf.

Bauman, Z. (1989), *Modernity and the Holocaust*, Cambridge: Polity Press.

Bauman, Z. (1993), *Postmodern Ethics*, Oxford: Blackwell Publishers.

Beatley, T. (1994), *Ethical Land Use: Principles of Policy and Planning*, Baltimore, MD: Johns Hopkins University Press.

Becker, L. C. and Becker, C. B. (eds) (1992), *Encyclopaedia of Ethics*, New York: Gerland Publishing.

Becker, L. C. and Kymlicka, W. (1995), 'Introduction', Symposium on Citizenship, Democracy and Education, *Ethics*, 105, pp. 465–7.

Beitz, C. R. (1991), 'Sovereignty and morality in international affairs', in Held (1991), pp. 236–54.

Bekker, S. (1997), 'Anthropology for Constitutional Court judges?' *Indicator South Africa*, 14(2), pp. 16–20.

Belsey, A. (1992), 'World poverty, justice and inequality', in Attfield and Watkins (1992), pp. 35–49.

Benhabib, S. (1992), *Situating the Self: Gender, Community and Postmodernism in Contemporary Ethics*, Cambridge: Polity Press.

Bianchini, F. (1990), 'The crisis of urban public social life in Britain: origins of the problem and possible responses', *Planning, Policy and Research*, 5(3), pp. 4–8.

Billington, R. (1993), *Living Philosophy: An Introduction to Moral Thought*, 2nd edition, London: Routledge.

Binnie, J. (1995), 'Trading places: consumption, sexuality and the production of queer space', in Bell, D. and Valentine, G. (eds), *Mapping Desires: Geographies of Sexuality*, London: Routledge, pp. 182–99.

Binnie, J. and Valentine, G. (1999), 'Geographies of sexuality: a review of progress', *Progress in Human Geography*, 23, pp. 175–87.

Birdsall, S. (1996), 'Regard, respect, and responsibility: sketches for a moral geography of the everyday', *Annals of the Association of American Geographers*, 86, pp. 619–29.

Black, R. (1996), 'Immigration and social justice: towards a progressive European immigration policy?' *Transactions of the Institute of British Geographers*, 21, pp. 64–75.

Block, G. and Drucker, M. (1992), *Rescuers: Portraits of Moral Courage in the Holocaust*, New York: Holmes & Meier.

Blumenfeld, J. (1997), 'From icon to scapegoat: the experience of South Africa's Reconstruction and Development Programme', *Development Policy Review*, 15, pp. 65–91.

Bond, P. (1998), 'Moving towards – or beyond? – a "post-Washington consensus" on development', *Indicator South Africa*, 15(4), pp. 30–41.

Booth, A. L. (1997), 'Critical questions in environmental philosophy', in Light and Smith (1997), pp. 255–73.

Boraine, A., Levy, J. and Scheffer, R. (eds) (1997), *Dealing with the Past: Truth and Reconciliation in South Africa*, 2nd edition, Cape Town: Institute for Democracy in South Africa.

Botman, H. R. and Petersen, R. M. (eds) (1996), *To Remember and to Heal: Theological and Psychological Reflections on Truth and Reconciliation*, Cape Town: Human & Rousseau.

Bowden, P. (1997), *Caring: Gender-sensitive Ethics*, London: Routledge.

Bowring, F. (1997), 'Communitarianism and morality: in search of the subject', *New Left Review*, 222, pp. 93–113.

Boyne, G. and Powell, M. (1991), 'Territorial justice: a review of theory and evidence', *Political Geography Quarterly*, 10, pp. 263–81.

Boyne, G. and Powell, M. (1993), 'Territorial justice and Thatcherism', *Environment and Planning C: Government and Policy*, 11, pp. 35–53.

Brennan, S. (1999), 'Recent work in feminist ethics', *Ethics*, 109, pp. 858–93.

Browett, J. (1980), 'Development and the diffusionist paradigm', *Progress in Human Geography*, 4, pp. 57–79.

Brown, A. (1986), *Modern Political Philosophy: Theories of the Just Society*, Harmondsworth: Penguin.

Brunn, S. (1989), 'Ethics in word and deed', *Professional Geographer*, 79, pp. iii-iv.

Buchanan, A. E. (1989), 'Assessing the communitarian critique of liberalism', *Ethics*, 99, pp. 852–82.

Buckle, S. (1993), 'Natural law', in P. Singer, (1991), pp. 161–74.

Bullard, R. (1990), *Dumping in Dixie: Race, Class, and Environmental Quality*, Boulder, CO: Westview Press.

Buttimer, A. (1974), *Values in Geography*, Commission on College Geography, Resource Paper 24, Washington, DC: Association of American Geographers.

Callicott, J. B. (1994), *Earth's Insight: A Multicultural Survey of Ecological Ethics from the Mediterranean Basin to the Australasian Outback*. Berkeley: University of California Press.

Carens, J. H. (1987), 'Aliens and citizens: the case for open borders', *The Review of Politics*, 49, pp. 251–73.

Chambers, R. (1997), 'Responsible well-being – a personal agenda for development', *World Development*, 25, 1743–54.

Charlesworth, A. (1994), 'Contesting places of memory: the case of Auschwitz', *Environment and Planning D: Society and Space*, 12, pp. 579–93.

Chisholm, M. and Smith, D. M. (eds) (1990), *Shared Space: Divided Space. Essays on Conflict and Territorial Organization*, London: Unwin Hyman.

Clark, G. L. (1986), 'Making moral landscapes: John Rawls' original position', *Political Geography Quarterly*, Supplement to 5, pp. 147–62.

Clement, G. (1996), *Care, Autonomy, and Justice: Feminism and the Ethic of Care*, Oxford: Westview Press.

Coates, B. E., Johnston, R. J. and Knox, P. L. (1977), *Geography and Inequality*, Oxford: Oxford University Press.

Cocking, D. and Oakly, J. (1995), 'Indirect consequentialism, friendship, and the problem of alienation', *Ethics*, 106, pp. 86–111.

Cohen, S. B. and Kliot, N. (1992), 'Place-names in Israel's ideological struggle over the administered territories', *Annals of the Association of American Geographers*, 82, pp. 653–80.

Cooper, D. E. and Palmer, J. A. (eds.) (1995), *Just Environments: Intergenerational, International and Interspecies Issues*, London: Routledge.

Corbridge, S. (1992), 'Development studies III: postcolonialism and development ethics', *Progress in Human Geography*, 16, pp. 322–34.

Corbridge, S. (1993a), 'Marxisms, modernities and moralities: development praxis and the claims of distant strangers', *Environment and Planning D: Society and Space*, 11, pp. 449–72.

Corbridge, S. (1993b), 'Development ethics and the international debt crisis', in Schuurman, F. (ed.), *Development Theories in the Nineties*, London: Zed Books, pp. 123–39.

Corbridge, S. (1994), 'Post-Marxism and post-colonialism: the needs and rights of distant strangers', in Booth, D. (ed.), *Rethinking Social Development: Theory, Research and Practice*, Harlow: Longman, pp. 90–117.

Corbridge, S. (1998), 'Development ethics: distance, difference, plausibility', *Ethics, Place and Environment*, 1, pp. 35–53.

Cornwell, J. (1984), *Hard-Earned Lives: Accounts of Health and Illness from East London*, London: Tavistock.

Cornwell, R. (1998), 'The African Renaissance: the state of the art', *Indicator South Africa*, 15(2), pp. 9–14.

Corry, M. R. (1991), 'On the possibility of ethics in geography: writing, citing and the construction of intellectual property', *Progress in Human Geography*, 15, pp. 125–47.

Cosgrove, D. (1989), 'Power and place in the Venetian territories', in Agnew, J. A. and Duncan, J. S. (eds), *The Power of Place: Bringing Together Geographical and Sociological Imagination*, London: Unwin Hyman, pp. 104–23.

Cosgrove, D. and Daniels, S. (1988), *The Iconography of Landscape*, Cambridge: Cambridge University Press.

Crampton, J. (1995), 'The ethics of GIS', *Cartography and Cartographic Information Systems*, 22, pp. 84–9.

Cresswell, T. (1994), 'Putting women in their place: the carnival at Greenham Common', *Antipode*, 26(1), pp. 35–58.

Cresswell, T. (1996), *In Place/Out of Place: Geography, Ideology and Transgression*, Minneapolis: University of Minnesota Press.

Crocker, D. (1991), 'Toward development ethics', *World Development*, 19, pp. 457–83.

Cuomo, C. J. (1998), *Feminism and Ecological Communities: An Ethic of Flourishing*, London: Routledge.

Cutter, S. L. (1995), 'Race, class and environmental justice', *Progress in Human Geography*, 19, pp. 111–22.

Cutter, S. L. and Solecki, W. D. (1996), 'Setting environmental justice in space and place: acute and chronic airborne toxic releases in the southeastern United States', *Urban Geography*, 17, pp. 380–99.

Dancy, J. (1993), *Moral Reasons*, Oxford: Blackwell Publishers.

Davies, W. K. D. and Herbert, D. (1993), *Communities within Cities: An Urban Social Geography*, London: Belhaven.

Dawson, A. H. (1979), 'Factories and cities in Poland', in French and Hamilton (1979), pp. 349–85.

Desai, A. (1999), *South Africa Still Revolting*, Johannesburg: Impact Africa Publishing.

de-Shalit, A. (1995), *Why Posterity Matters: Environmental Policies and Future Generations*, London: Routledge.

Dobroszycki, L. (1984), *The Chronicle of the Łódź Ghetto 1941–1944*, New Haven and London: Yale University Press.

Domański, B. (1992), 'Social control over the milltown: industrial paternalism under socialism and capitalism', *Tijdschrift voor Economische en Sociale Geografie*, 83, pp. 353–60.

Domański, B. (1997), *Industrial Control over the Socialist Town: Benevolence or Exploitation?* Westport, CO, and London: Praeger.

Douglas, M. and Friedmann, J. (eds) (1998), *Cities for Citizens: Planning and the Rise of Civil Society in a Global Age*, Chichester: John Wiley.

Dower, N. (1989), *What is Development: A Philosopher's Answer*, Occasional Paper 3, Glasgow: Centre for Development Studies, University of Glasgow.

Dower, N. (1992), 'Sustainability and the right to development', in Attfield and Wilkins (1992), pp. 93–116.

Dowling, R. (1998), 'Suburban stories, gendered lives: thinking through difference', in Fincher, R. and Jacobs, J. M. (eds), *Cities of Difference*, New York and London: The Guildford Press, pp. 69–88.

Doyal, L. and Gough, I. (1991), *A Theory of Human Need*, London: Macmillan.

Doyle, T. (1998), 'Sustainable development and Agenda 21: the secular bible of global free markets and pluralist democracy', *Third World Quarterly*, 19, pp. 771–86.

Driver, F. (1988), 'Moral geographies: social science and the urban environment in mid-nineteenth century England', *Transactions of the Institute of British Geographers*, 13, pp. 275–87.

Driver, F. (1991), 'Morality, politics, geography: brave new words', in Philo (1991), pp. 61–4.

Dunn, J. (1984), *John Locke*, Oxford: Oxford University Press.

du Toit, A. (1997), 'No rest without the wicked: assessing the Truth Commission', *Indicator South Africa*, 14(1), pp. 7–12.

Dwivedi, O. P. (1990), '*Satyagraha* for conservation: awakening the spirit of Hinduism', in Engel and Engel (1990), pp. 201–12.

Eagleton, T. (1996), *The Illusions of Postmodernism*, Oxford: Blackwell Publishers.

Eichengreen, L. (1994), *From Ashes to Life: My Memories of the Holocaust*, San Francisco: Mercury House.

Elon, A. (1996), *Jerusalem: City of Mirrors*, London: Flamingo.

Engel, J. R. 'Introduction: The ethics of sustainable development', in Engel and Engel (1990), pp. 1–23.

Engel, J. R. and Engel, J. G. (eds) (1990), *Ethics of Environment and Development: Global Challenge, International Response*, London: Belhaven.

Escobar, A. (1996), 'Constructing nature: elements for a poststructural political ecology', in Peet and Watts (1996), pp. 46–68.

Etzioni, A. (1988), *The Moral Dimension: Towards a New Economics*, New York: Free Press.

Etzioni, A. (1995), *The Spirit of Community: Rights, Responsibilities and the Communitarian Agenda*, London: Fontana Press.

Fair, T. J. D. (1982), *South Africa: Spatial Frameworks for Development*, Cape Town: Juta & Co.

Fein, H. (1979), *Accounting for Genocide: National Responses and Jewish Victimization During the Holocaust*, Chicago: University of Chicago Press.

Fine, B. (1995), 'Privatisation and the RDP: a critical assessment', *Transformation*, 27, pp. 1–23.

Fine, R. and van Wyk, G. (1996), 'South Africa: state, labour, and the politics of reconstruction', *Capital and Class*, 58, pp. 19–31.

Fisk, M. (1995), 'Justice and universality', in Sterba et al. (1995), pp. 221–44.

Fitton, R. S. and Wadsworth, A. P. (1958), *The Strutts and the Arkwrights, 1758–1830: A Study of the Early Factory System*, Manchester: Manchester University Press.

Fogelman, E. (1994), *Conscience and Courage: Rescuers of Jews During the Holocaust*, New York: Anchor Books.

Foster-Carter, A. (1976), 'From Rostow to Gunder Frank: conflicting paradigms in the analysis of underdevelopment', *World Development*, 4, pp. 167–180.

Frank, A. G. (1971), *Capitalism and Underdevelopment in Latin America*, Harmondsworth: Penguin.

Fraser, N. (1995), 'From redistribution to recognition? Dilemmas of justice in a "post-socialist" age', *New Left Review*, 212, pp. 68–93.

French, R. A. and Hamilton, F. E. I. (eds) (1979), *The Socialist City: Spatial Structure and Urban Policy*, Chichester: John Wiley.

Fried, C. (1983), 'Distributive justice', *Social Philosophy and Policy*, 1, pp. 45–59.

Friedman, M. (1991a) 'Feminism and modern friendship: dislocating the community', in Arthur and Shaw (1991), pp. 304–19.

Friedman, M. (1991b), 'The practice of partiality', *Ethics*, 101, pp. 818–35.

Friedman, M. (1993), *What are Friends for? Feminist Perspectives on Personal Relationships and Moral Theory*, Ithica and London: Cornell University Press.

Friedmann, J. (1966), *Regional Development Policy*, Cambridge, MA: MIT Press.

Friedmann, J. (1992), *Empowerment: The Politics of Alternative Development*, Oxford: Blackwell Publishers.

Friedmann, J. (1996), 'Borders, margins and frontiers: myth and metaphor', in Gradus, Y. and Lithwick, H. (eds), *Frontiers in Regional Development*, London: Rowman & Littlefield, pp. 1–20.

Geertz, C. (1973), *The Interpretation of Cultures*, New York: Basic Books.

Gensler, H. J. (1998), *Ethics: A Contemporary Introduction*, London: Routledge.

Gensler, W. M. (1992), 'Therapeutic landscapes: medical geographic research in the light of the new cultural geography', *Social Science and Medicine*, 34, pp. 735–46.

Geras, N. (1995), *Solidarity in the Conversation of Humankind: The Ungroundable Liberalism of Richard Rorty*, London: Verso.

Gerber, J. (1997), 'Beyond dualism – the social construction of nature *and* social construction of human beings', *Progress in Human Geography*, 21, pp. 1–17.

Geyer-Ryan, H. (1996), 'From morality to mortality: women and the violence of political change, or law and (b)order', *Philosophy & Social Criticism,* 22(4), pp. 1–11.

Giddens, A. (1979), *Central Problems in Social Theory*, London: Macmillan.

Gilligan, C. (1982), *In a Different Voice: Psychological Theory and Women's Development*, Cambridge, MA: Harvard University Press.

Gilligan, C. (1987), 'Moral orientation and moral development', in Kittay, E. and Meyers, D. (eds), *Women and Moral Theory*, Totowa, NJ: Rowman & Littlefield, pp. 19–33.

Ginzburg, C. (1994), 'Killing a Chinese mandarin: the moral implications of distance', *New Left Review*, 208, pp. 107–20.

Gleeson, B. (1996), 'Justifying justice', *Area*, 28, pp. 229–34.

Glinert, L. and Shilhav, Y. (1991), 'Holy land, holy language: a study of an Ultraorthodox Jewish ideology', *Language in Society*, 20, pp. 59–86.

Goldberg, D. J. and Rayner, J. D. (1989), *The Jewish People: Their History and Their Religion*, Harmondsworth: Penguin.

Goldhagen, D. J. (1996), *Hitler's Willing Executioners: Ordinary Germans and the Holocaust,* London: Little, Brown.

Goodin, R. E. (1991), 'Utility and the good', in P. Singer (1991), pp. 241–8.

Goulet, D. (1990), 'Development ethics and ecological wisdom', in Engel and Engel (1990), pp. 36–49.

Goulet, D. (1995), *Development Ethics: A Guide to Theory and Practice*, Notre Dame: Notre Dame University Press.

Gower, B. S. (1995), 'The environment and justice for future generations', in Cooper and Palmer (1995), pp. 49–58.

Gradus, Y. and Lipshitz, G. (eds) (1996), *The Mosaic of Israeli Geography*, Beer Sheva: Ben-Gurion University of the Negev Press.

Graham, G. (1990), *Living the Good Life: An Introduction to Moral Philosophy*, New York: Paragon House.

Graham, G. (1997), *Ethics and International Relations*, Oxford: Blackwell Publishers.

Greaves, D. (1994), 'Marx, justice and history', *Theoria*, 83/84, pp. 13–35.

Griffin, J. (1986), *Well-being: Its Meaning, Measurement and Moral Importance,* Oxford: Clarendon Press.

Grimes, S. (1999), 'Exploring the ethics of development', in Proctor and Smith (1999), pp. 59–71.

Grodsky, B. (1997), 'City Council rejects plan to rebuild Warsaw's largest square', *Warsaw Business Journal,* 20–26 October, p. 5.

Gruen, L. (1991), 'Animals', in P. Singer (1991), pp. 343– 53.

Habermas, J. (1990), *Moral Consciousness and Communicative Action,* Cambridge: Polity Press.

Hansen, C. (1991), 'Classical Chinese ethics', in P. Singer (1991), pp. 69–81.

Harcourt, W. (1997), 'The search for social justice', *Development*, 40, pp. 5–11.

Harding, S. (1987), 'The curious coincidence of feminine and African moralities', in Kittay, E. and Myers, D. T. (eds), *Women and Moral Theory*, Totowa, NJ: Rowman & Littlefield, pp. 296–315; as reprinted in Eze, E. C. (ed.) (1998), *African Philosophy: An Anthology*, Oxford: Blackwell Publishers, pp. 360–72.

Harman, G. and Thompson, J. J. (1996), *Moral Relativism and Moral Objectivity*, Oxford: Blackwell Publishers.

Harvey, D. (1972), 'Social justice and spatial systems', in Peet, R. (ed.), *Geographical Perspectives on American Poverty*, Antipode Monographs on Social Geography, 1, Worcester, MA: pp. 87–106.

Harvey, D. (1973), *Social Justice and the City*, London: Edward Arnold.

Harvey, D. (1979), 'Monument and myth', *Annals of the Association of American Geographers*, 69, pp. 362–81.

Harvey, D. (1996), *Justice, Nature and the Geography of Difference*, Oxford: Blackwell Publishers.

Hasson, S. (1996), *The Cultural Struggle over Jerusalem: Accommodations, Scenarios and Lessons*, Jerusalem: The Floersheimer Institute for Policy Studies.

Hasson, S. and Gonen, A. (1997), *The Cultural Tension within Jerusalem's Jewish Population*, Jerusalem: The Floersheimer Institute for Policy Studies.

Hay, A. M. (1995), 'Concepts of equity, fairness and justice in geographical studies', *Transactions of the Institute of British Geographers*, 20, pp. 500–8.

Hay, I. (1998), 'Making moral imaginations: research ethics, pedagogy, and professional human geography', *Ethics, Place and Environment*, 1, pp. 55–75.

Heiman, M. K. (1996), 'Race, waste, and class: new perspectives on environmental justice, *Antipode*, 28, pp. 111–21.

Hekman, S. (1995), *Moral Voices, Moral Selves: Carol Gilligan and Feminist Moral Theory*, Cambridge: Polity Press.

Held, D. (ed.) (1991), *Political Theory Today*, Cambridge: Polity Press.

Held, V. (1993), *Feminist Morality: Transforming Culture, Society, and Politics*, Chicago and London: University of Chicago Press.

Heller, A. (1987), *Beyond Justice*, Oxford: Blackwell Publishers.

Hendler, S. (ed.) (1995), *Planning Ethics: A Reader in Planning Theory, Practice, and Education*, New Brunswick, NJ: Center for Urban Policy Research, Rutgers University.

Hillary, G. (1955), 'Definitions of community: areas of agreement', *Rural Sociology*, 20, pp. 111–23

Hillier, J. (1998), 'Paradise proclaimed? Towards a theoretical understanding of representations of nature in land use planning decision-making', *Ethics, Place and Environment*, 1, pp. 77–91.

Hinman, L. M. (1994), *Ethics: A Pluralistic Approach to Moral Theory*, Fort Worth: Harcourt Brace College Publishers.

Hirst, P. and Thompson, G. (1995), 'Globalization and the future of the nation state', *Economy and Society*, 24, pp. 408–42.

Hoffman, E. (1998), *Shtetl: The Life and Death of a Small Town and the World of Polish Jews*, London: Secker & Warburg.

Holloway, S. H. (1998), 'Local childcare cultures: moral geographies of mothering and the social organisation of pre-school education', *Gender, Place and Culture*, 5, pp. 29–53.

Houston, J. M. (1978), 'The concepts of "place" and "land" in the Judaeo-Christian tradition', in Ley, D. and Samuels, M. (eds), *Humanistic Geography: Prospects and Problems*, Chicago: Maaroufa, pp. 224–37.

Howe, E. (1994), *Acting on Ethics in City Planning*, New Brunswick, NJ: Centre for Urban Policy Research, Rutgers University.

Howell, S. (ed.) (1997), *The Ethnography of Moralities*, London: Routledge.

Hubbard, P. (1997), 'Red-light districts and toleration zones: geographies of female street prostitution in England and Wales', *Area*, 29, pp. 129–40.

Hubbard, P. (1998), 'Sexuality, immorality and the city: red-light districts and the marginalisation of female street prostitutes', *Gender, Place and Culture*, 5, pp. 55–72.

Humphrey, C. (1997), 'Exemplars and rules: aspects of the discourse of moralities in Mongolia', in Howell (1997), pp. 25–47.

Huyssen, A. (1990), 'Mapping the postmodern', in Nicholson (1990), pp. 234–77.

Ignatieff, M. (1984), *The Needs of Strangers*, London: Chatto & Windus.

Israel Wire, www.israel.wire.com/

Jabłoński, S. and Jabłoński, K. (1995), *Łódź*, Warsaw: Festina.

Jackson, F. (1991), 'Decision-theoretic consequentialism and the nearest and dearest objection', *Ethics*, 101, pp. 461–82.

Jackson, P. (1984), 'Social disorganization and moral order in the city', *Transactions of the Institute of British Geographers*, 9, pp. 168–80.

Jackson, P. (1989), *Maps of Meaning: An Introduction to Cultural Geography*, London: Unwin Hyman.

Jackson, P. and Smith, S. J. (1984), *Exploring Social Geography*, London: Allen & Unwin.

Jacobs, H. M. (1995), 'Contemporary environmental philosophy and its challenge to planning theory', in Hendler (1995), pp. 83–103.

Jacobson-Widding, A. (1997), '"I lied, I farted, I stole . . . ": dignity and morality in African discourse on personhood', in Howell (1997), pp. 48–73.

Jagger, A. M. (1995) 'Toward a feminist conception of moral reasoning', in Sterba et al (1995), pp. 115–46.

Jeffery, A. (1997), *Bill of Rights Report 1996/97*, Johannesburg: South African Institute of Race Relations.

Jeffery, A. (1999), *The Truth about the Truth Commission*, Johannesburg: South African Institute of Race Relations.

Johnston, R. J. (1999), 'Geography, fairness, and liberal democracy', in Proctor and Smith (1999), pp. 44–58.

Johnston, R. J., Gregory, D. and Smith, D. M. (eds) (1994), *The Dictionary of Human Geography*, 3rd Edition, Oxford: Blackwell Publishers.

Jones, P. (1994), *Rights*, London: Macmillan.

Kaczmarek, S. (1995), 'Urban space, living conditions and residential preference in *Łódź*, in Smith, D. M. (ed.) (1995), *Łódź: Geographical Studies of a Polish City*, Research Paper 8, London: Department of Geography, Queen Mary and Westfield College, pp. 9–29.

Kagan, J. (1992), *Polish Jewish Heritage*, New York: Hippocrene Books.

Karpf, A. (1996), *The War After: Living with the Holocaust*, London: Heinemann.

Kekes, J. (1994), 'Pluralism and the value of life', in Paul, Miller and Paul (1994), pp. 44–60.

Kellner, M. (1991), 'Jewish ethics', in P. Singer (1991), pp. 82–90.

Kimmerling. B. (1983), *Zionism and Territory: The Socio-territorial Dimensions of Zionist Politics*, Berkeley: Institute of International Studies, University of California.

Kimmerling. B. and Migdal, J. S. (1994), *Palestinians: The Making of a People*, Cambridge, MA: Harvard University Press.

King, R. (1997), 'Critical reflections on biocentric environmental ethics: is it an alternative to anthropocentrism?' in Light and Smith (1997), pp. 209–30.

Kirby, A. (1991), 'On ethics and power in higher education', *Journal of Geography in Higher Education*, 15, pp. 75–7.

Kitchen, R. M. (1998), 'Towards geographies of cyberspace', *Progress in Human Geography*, 22, pp. 385–406.

Knox, P. L. (1995), *Social Well-being: A Spatial Perspective*, Oxford: Oxford University Press.

Kobayashi, A. (1997), 'The paradox of difference and diversity (or, why the threshold keeps moving)', in Jones, J. P., Nast, H. J. and Roberts, S. M. (eds) *Thresholds in Feminist Geography: Difference, Methodology, Representation*, Oxford: Rowman & Littlefield, pp. 3–9.

Kohlberg, L. (1981), *The Philosophy of Moral Development*, San Francisco: Harper & Row.

Kothari, R. (1990), 'Environment, technology, and ethics', in Engel and Engel (1990), pp. 27–35.

Kuehls, T. (1996), *Beyond Sovereign Territories: The Space of Ecopolitics*, Minneapolis: University of Minnesota Press.

Kukathas, C. (1994), 'Explaining moral variety', in Paul, Miller and Paul (1994), pp. 1–21.

Kukathas, C. (1997), 'Cultural toleration', in Kymlicka and Shapiro (1997), pp. 69–104.

Kymlicka, W. (1990), *Contemporary Political Philosophy: An Introduction*, Oxford: Clarendon Press.

Kymlicka, W. (1993), 'Community', in Goodin, R. E. and Pettit, P. (eds), *A Companion to Contemporary Political Philosophy*, Oxford: Blackwell Publishers, pp. 366–78.

Kymlicka, W. (1995), *Multicultural Citizenship: A Liberal Theory of Minority Rights*, Oxford: Clarendon Press.

Kymlicka, W. and Shapiro, I. (eds) (1997), *Ethnicity and Group Rights*, New York: New York University Press.

Lafollette, H. (1991), 'Personal relationships', in P. Singer (1991), pp. 327–32.

Lake, R. W. (1993), 'Planning and applied geography: positivism, ethics, and geographical information systems', *Progress in Human Geography*, 17, pp. 404–13.

Lake, R. W. (1996), 'Volunteers, NIMBYs, and environmental justice: dilemmas of democratic practice', *Antipode*, 28, pp. 160–74.

Langhelle, O. (1999), 'Sustainable development: exploring the ethics of *Our Common Future*', *International Political Science Review*, 20, pp. 129–49.

Laws, G. (ed.) (1994), 'Special issue: social (in)justice in the city: theory and practice two decades later', *Urban Geography*, 15(7).

Le Grand, J. (1991), *Ethics and Choice: An Essay in Economics and Applied Philosophy*, London: HarperCollins.

Lee, R. (1996), 'Moral money? LETS and the social construction of local economic geographies in Southeast England', *Environment and Planning A*, 28, pp. 1377–94.

Leopold, A. (1949), *A Sand County Almanac*, reprint 1970, San Francisco and New York: Sierra Club/Ballantine Books; chapter entitled 'The land ethic' reprinted in Agnew, J., Livingstone, D. N. and Rodgers, A. (eds) (1996), *Human Geography: An Essential Anthology*, Oxford: Blackwell Publishers, pp. 351–64.

Ley, D. (1987), 'Styles of the times: liberal and neo-conservative landscapes in inner Vancouver, 1968–86', *Journal of Historical Geography*, 13, pp. 40–56.

Ley, D. (1993), 'Co-operative housing as a moral landscape: re-examining "the post-modern city"', in Duncan, J. and Ley, D. (eds), *Place/Culture/Representation*, London: Routledge, pp. 128–48.

Light, A. (1996), 'Callicott and Naess on pluralism', *Inquiry*, 39, pp. 273–94.

Light, A. and Smith, J. M. (eds) (1997), *Space, Place, and Environmental Ethics: Philosophy and Geography I*, London: Rowman & Littlefield.

Liszewski, S. (1991), 'The role of the Jewish community in the organiszation of urban space in Łódź', *Polin: A Journal of Polish-Jewish Studies*, 6, pp. 27–36.

Liszewski, S. (1997), 'The origins and stages of development of industrial Łódź and the Łódź urban region', in Liszewski and Young (1997), pp. 11–34.

Liszewski, S. and Young, C. (eds) (1997), *A Comparative Study of Łódź and Manchester: Geographies of European Cities in Transition*, Łódź: Łódź University Press.

Livingstone, D. N. (1991), 'The moral discourse of climate: Historical considerations on race, place and virtue', *Journal of Historical Geography*, 17, pp. 413–34.

Livingstone, D. N. (1992), *The Geographical Tradition*, Oxford: Blackwell Publishers.

Livingstone, D. N. (1995), 'The polity of nature: representation, virtue, strategy', *Ecumene*, 2, pp. 353–77.

Louden, R. B. (1992), *Morality and Moral Theory: A Reappraisal and Reaffirmation*, Oxford: Oxford University Press.

Low, N. and Gleeson, B. (1998), *Justice, Society and Nature: An Exploration of Political Ecology*, London: Routledge.

Low, N. and Gleeson, B. (1999), 'Geography, justice and the limits of rights', in Proctor and Smith (1999), pp. 30–43.

Lukas, R. (1997), *Forgotten Holocaust: The Poles under German Occupation*, 2nd edition, New York: Hippocrene Books.

Lynn, W. S. (1998), 'Contesting moralities: animals and moral values in the Dear/Symanski debate', *Ethics, Place and Environment*, 1, pp. 223–42.

McDowell, L. (1994), 'Polyphony and pedagogic authority', *Area*, 26, pp. 41–8.

McDowell, L. (1995), 'Understanding diversity: the problem of/for theory', in Johnston, R. J., Taylor, P. J. and Watts, M. J. (eds), *Geographies of Global Change: Remapping the World in the Ttwentieth Century*, Oxford: Blackwell Publishers, pp. 280–94.

MacIntyre, A. (1985), *After Virtue: A Study in Moral Theory*, 2nd edition, London: Duckworth.

MacIntyre, A. (1998), *A Short History of Ethics: A History of Moral Philosophy from the Homeric Age to the Twentieth Century*, 2nd edition, London: Routledge.

Mackie, J. L. (1977), *Ethics: Inventing Right and Wrong*, Harmondsworth: Penguin.

Mandela, N. (1994), *Long Walk to Freedom*, London: Little, Brown.

Marais, H. (1998), *South Africa: Limits to Change: The Political Economy of Transition*, London and New York: Zed Books; Cape Town: UCT Press.

Margalit, A. and Halbertal, M. (1994), 'Liberalism and the right to culture', *Social Research*, 61, pp. 491–510.

Markovic, M. (1990), 'The development vision of socialist humanism', in Engel and Engel (1990), pp. 127–36.

Marris, R. (1974), *Loss and Change*, London: Routledge & Kegan Paul.

Martin, J. R. (1996) 'Aerial distance, esotericism, and other closely related traps', *Signs: Journal of Women in Culture and Society*, 21, pp. 584–614.

Massey, D. (1991), 'A global sense of place', *Marxism Today*, June, pp. 24–9.

Matchen, T. R. (1995), 'Justice, self, and natural rights', in Sterba et al. (1995), pp. 59–106.

Matless, D. (1994), 'Moral geographies in Broadlands', *Ecumene*, 1, pp. 27–56.

Matless, D. (1995), 'Culture run riot? Work in social and cultural geography, 1994', *Progress in Human Geography*, 19, pp. 395–403.

Matless, D. (1997), 'Moral geographies of English landscape', *Landscape Research*, 22, pp. 141–55.

Matthews, E. (1994), 'A post-modern man's pat answer', *The Times Higher Educational Supplement*, 3 June.

Mbigi, L. and Maree, J. (1995), *Ubuntu: The Spirit of African Transformation Management*, Randburg: Knowledge Resources.

Mendus, S. (1993), 'Different voices, still lives: problems in the ethics of care', *Journal of Applied Philosophy*, 10, pp. 17–27.

Merchant, C. (1996), *Earthcare: Women and the Environment*, London: Routledge.

Midgley, M. (1991), 'The origin of ethics', in P. Singer (1991), pp. 3–13.

Milgram, S. (1974), *Obedience to Authority: An Experimental View*, London: Tavistock.

Miller, R. W. (1992), *Moral Differences: Truth, Justice and Conscience in a World of Conflict*, Princeton, NJ: Princeton University Press.

Minchinton, W. (1984), *A Guide to Industrial Archaeology Sites in Britain*, London: Grenada Publishing.

Mitchell, B. and Draper, D. (1982), *Relevance and Ethics in Geography*, London: Longman.

Morris, W. C. (1996), 'Well-being, reason, and the politics of law', *Ethics*, 106, pp. 817–33 [review of Raz, J. (1994), *Ethics in the Public Domain: Essays in the Morality of Law and Politics*, Oxford: Clarendon Press].

Mosher, A. E. (1995), '"Something better than the best": industrial restructuring, George McMurtry and the creation of the model industrial town of Vandergrift, Pennsylvania, 1883–1901', *Annals of the Association of American Geographers*, 85, pp. 84–107.

Mulhall, S. and Swift, A. (1996), *Liberals and Communitarians*, 2nd edition, Oxford: Blackwell Publishers.

Naess, A. (1989), *Ecology, Community and Lifestyle*, translated by D. Rothenburg, Cambridge: Cambridge University Press.

Nagel, T. (1986), *The View from Nowhere*, Oxford: Oxford University Press.

Nagel, T. (1993), 'Personal rights and public space', *Philosophy and Public Affairs*, 24, pp. 83–107.

Nanji, A. (1991), 'Islamic ethics', in P. Singer (1991), pp. 106–18.

Nattrass, N. (1996), 'Gambling on investment: competing economic strategies in South Africa', *Transformation*, 31, pp. 25–42.

Nett, R. (1971), 'The civil right we are not ready for: the right of free movement of people on the face of the Earth', *Ethics*, 81, pp. 212–27.

Neuberger, B. (1997), *Religion and Democracy in Israel*, Jerusalem: The Floersheimer Institute for Policy Studies.

Neuberger, J. (1996), *On Being Jewish*, London: Mandarin.

Newman, D. (1984), 'Ideological and political influences on Israeli urban colonization: the West Bank and Galilee Mountains', *Canadian Geographer*, 28 (2), pp. 142–55.

Newman, D. (ed.) (1985), *The Impact of Gush Emunim: Politics and Settlement in the West Bank*, New York: St Martin's Press.

Newman, D. (1999), 'A new agenda', *Jerusalem Post*, 19 May.

Newman, D. and Paasi, A. (1998), 'Fences and neighbours in the postmodern world: boundary narratives in political geography', *Progress in Human Geography*, 22, pp. 186–207.

Nichloson, L. J. (ed.) (1990), *Feminism/Postmodernism*, London: Routledge.

Noddings, N. (1984), *Caring: A Feminine Approach to Ethics and Moral Education*, Berkeley: University of California Press.

Nozick, R. (1974), *Anarchy, State, and Utopia*, New York: Basic Books.

Ntuli, P. P. (1998), 'Who's afraid of the African Renaissance?' *Indicator South Africa*, 15(2), pp. 15–18.

Nussbaum, M. C. (1992), 'Human functioning and social justice: in defence of Aristotelian essentialism', *Political Theory*, 20, pp. 202–46.

Nussbaum, M. C. (1998), 'Public philosophy and international feminism', *Ethics*, 108, pp. 762–96.

Nussbaum, M. C. and Glover, J. (eds) (1995), *Women, Culture and Development: A Study of Human Capabilities*, Oxford: Clarendon Press.

O'Donovan, M. (1995), 'Indicating development: the Human Development Index', *Indicator South Africa*, 12(3), pp. 91–5.

Offer, A. (1997), 'Between the gift and the market: the economy of regard', *Economic History Review*, 50, pp. 450–76.

Ogborn, M. and Philo, C. (1994), 'Soldiers, sailors and moral locations in nineteenth-century Portsmouth', *Area*, 26(3), pp. 221–31.

Okin, S. M. (1989), 'Reason and feeling in thinking about justice', *Ethics*, 99, pp. 229–49.

Okin, S. M. (1998), 'Feminism and muticulturalism: some tensions', *Ethics*, 108, pp. 661–84.

Oliner, S. P. and Oliner, P. M. (1992), *The Altruistic Personality: Rescuers of Jews in Nazi Europe*, New York: The Free Press.

Omari, C. K. (1990) 'Traditional African land ethics', in Engel and Engel (1990), pp. 167–75.

Omo-Fadaka, J. (1990), 'Communalism: The moral factor in African development', in Engel and Engel (1990), pp. 176–82.

O'Neil, O. (1986), *Faces of Hunger: An Essay on Poverty, Justice and Development*, London: Allen & Unwin.

O'Neill, O. (1991), 'Transnational justice', in Held (1991), pp. 277–304.

O'Neill, O. (1992), 'Justice, gender and international boundaries', in Attfield and Wilkins (1992), pp. 50–76.

O'Neill, O. (1996), *Towards Justice and Virtue: A Constructive Account of Practical Reasoning*, Cambridge: Cambridge University Press.

Oosthuizen, G. (1996), 'African independent/indigenous churches in the social environment: an empirical analysis', *Africa Insight*, 26, pp. 308–24.

Ó Tuathail, G. (1996), 'Political geography II: (counter) revolutionary times', *Progress in Human Geography*, 20, pp. 404–12.

Palmer, M. (1990), 'The encounter of religion and conservation', in Engel and Engel (1990), pp. 50–62.

Paul, E. F., Miller, F. D. and Paul, J. (eds) (1994), *Cultural Pluralism and Moral Knowledge*, Cambridge: Cambridge University Press.

Peet, R. (1977), *Radical Geography: Alternative Viewpoints on Contemporary Social Issues*, London: Methuen.

Peet, R and Watts, M. (eds) (1996), *Liberation Ecologies: Environment, Development, Social Movements*, London: Routledge.

Peffer, R. G. (1990), *Marxism, Morality, and Social Justice*, Princeton, NJ: Princeton University Press.

Pepper, D. (1993), *Eco-socialism: From Deep Ecology to Social Justice*, London: Routledge.

Philo, C. (compiler) (1991), *New Words, New Worlds: Reconceptualising Social and Cultural Geography*, Lampeter: Department of Geography, St David's University College.

Pietilä, H. (1990), 'The daughters of earth: women's culture as a basis for sustainable development', in Engel and Engel (1990), pp. 235–44.

Ploszajska, T. (1994), 'Moral landscapes and manipulated spaces: gender, class and space in Victorian reformatory schools', *Journal of Historical Geography*, 20, pp. 413–29.

Pogge, T. W. (1992), 'Cosmopolitanism and sovereignty', *Ethics*, 103, pp. 48–75.

Polansky, A. (1989), 'Polish-Jewish relations and the Holocaust', *Polin: A Journal of Polish-Jewish Studies*, 4, pp. 226–42.

Poole, R. (1991), *Morality and Modernity*, London: Routledge.

Popławska, I. and Muthesis, S. (1986), 'Poland's Manchester: 19th-century industrial and domestic architecture in Łódź', *Journal of the Society of Architectural Historians*, 45, pp. 149–60.

Posner, R. A. (1998), 'The problematics of moral and legal theory', *Harvard Law Review*, 111, pp. 1637–717.

Powers, C. T. (1997), *In the Memory of the Forest*, London: Anchor.

Preston, R. (1991), 'Christian ethics', in P. Singer (1991), pp. 91–105.

Price, R. (1997), 'Race and reconciliation in the New South Africa', *Politics & Society*, 25, pp. 149–78.

Proctor, J. D. (1995), 'Whose nature? The contested moral terrain of ancient forests', in Cronon, W. (ed.), *Uncommon Ground: Towards Reinventing Nature*, New York: W. W. Norton, pp. 269–97.

Proctor, J. D. (1998a), 'Ethics in geography: giving moral form to the geographical imagination', *Area*, 30, pp. 8–18.

Proctor, J. D. (1998b), 'Expanding the scope of science and ethics', *Annals of the Association of American Geographers*, 88, pp. 290–6.

Proctor, J. D. (1998c), 'The social construction of nature: relativist accusations, pragmatist and critical realist responses', *Annals of the Association of American Geographers*, 88, pp. 352–76.

Proctor, J. D. (1998d), 'Geography, paradox and environmental ethics', *Progress in Human Geography*, 22, pp. 235–55.

Proctor, J. D. (1999), 'The spotted owl and the contested moral landscape of the Pacific Northwest', in Wolch, J. and Emel, J. (eds), *Animal Geographies*, London: Verso, pp. 191–217.

Proctor, J. D. and Smith, D. M. (eds) (1999), *Geography and Ethics: Journeys in a Moral Terrain*, London: Routledge.

Prokopówna, E. (1989), 'The image of the *shtetl* in Polish literature', *Polin: A Journal of Polish-Jewish Studies*, 4, pp. 129–42.

Pulido, L. (1996), 'A critical review of the methodology of environmental racism research', *Antipode*, 28(2), pp. 142–9.

Puś, W. (1991), 'The development of the city of Łódź (1820–1939)', *Polin: A Journal of Polish-Jewish Studies*, 6, pp. 3–19.

Rainbow, P. (1991), *The Foucault Reader*, Harmondsworth: Penguin.

Rapport, N. (1997), 'The morality of locality: on the absolutism of landownership in an English village', in Howell (1997), pp. 74–97.

Rauche, G. V. (1985), *Theory and Practice in Philosophical Argument*, Durban: Institute for Social and Economic Research, University of Durban-Westville.

Rawls, J. (1971), *A Theory of Justice*, Cambridge, MA: Harvard University Press.

Reagan, D. (1996), 'The land of Israel: to whom does it belong?', Lamb and Lion Ministries: http://www.lamblion.com/

Reiman, J. (1990), *Justice and Modern Moral Philosophy*, New Haven: Yale University Press.

Rheingold, H. (1993), *The Virtual Community: Homesteading on the Electronic Frontier*, Reading, MA: Addison-Wesley.

Riley, R. (1997), 'Central area activities in a post-communist city: Łódź, Poland', *Urban Studies*, 34, pp. 453–70.

Riley, R. (1998), 'Łódź textile mills: indigenous culture or functional imperatives?' *Industrial Archaeology Review*, 20, pp. 91–104.

Robinson, D. and Garratt, C. (1996), *Ethics for Beginners*, Cambridge: Icon Books.

Robinson, J. (1997), 'The geopolitics of South African cities', *Political Geography*, 16, pp. 365–86.

Roemer, J. E. (1996), *Theories of Distributive Justice*, Cambridge, MA: Harvard University Press.

Rolston, H. (1993), 'Rights and responsibilities on the home planet', *The Yale Journal of International Law*, 18, pp. 251–79.

Rorty, R. (1989), *Contingency, Irony, and Solidarity*, Cambridge: Cambridge University Press.

Rose, G. (1997), 'Situating knowledge: positionality, reflectivities and other tactics', *Progress in Hyman Geography*, 21, pp. 305–20.

Rostow, W. W. (1960), *The Sages of Economic Growth: A Non-Communist Manifesto*, Cambridge: Cambridge University Press.

Rowe, C. (1991), 'Ethics in ancient Greece', in P. Singer (1991), pp. 121–32.

RSA (1994), *White Paper on Reconstruction and Development: Government's Strategy for Fundamental Transformation*, Pretoria.

RSA (1996), *Growth, Employment and Redistribution: A Macroeconomic Strategy*, Pretoria.

Rubinstein, D. (1991), *The People of Nowhere*, New York: Random House.

Sack, R. D. (1986), *Human Territoriality: Its Theory and History*, Cambridge: Cambridge University Press.

Sack, R. D. (1997), *Homo Geographicus: A Framework for Action, Awareness, and Moral Concern*, Baltimore and London: The Johns Hopkins University Press.

Sandel, M. (1982), *Liberalism and the Limits of Justice*, Cambridge: Cambridge University Press.

Sandercock, L. (1998), *Towards Cosmopolis: Planning for Multicultural Cities*, Chichester: John Wiley.

Sayer, A. and Storper, M. (1997), 'Ethics unbound: for a normative turn in social theory', *Environment and Planning D: Society and Space*, 15, pp. 11–17.

Sayigh, R. (1979), *Palestinians: From Peasants to Revolutionaries*, London: Zed Books.

Schlemmer, L. and Levitz, C. (1998), *Unemployment in South Africa: The Facts, the Prospects, and an Exploration of Solutions*, Johannesburg: South African Institute of Race Relations.

Selznick, P. (1992), *The Moral Commonwealth: Social Theory and the Promise of Community*, Berkeley: University of California Press.

Sen, A. (1987), *On Ethics and Economics*, Oxford: Basic Books.

Sen, A. (1992), *Inequality Reexamined*, Oxford: Clarendon Press.

Shapiro, M. J. (1994), 'Moral geographies and the ethics of post-sovereignty', *Public Culture*, 6, pp. 479–502.

Shilhav, Y. (1983), 'Communal conflict in Jerusalem – the spread of ultra-orthodox neighborhoods', in Kliot, N. and Waterman, S. (eds), *Pluralism and Political Geography: People, Territory and the State*, London: Croom Helm, pp. 100–13.

Shilvav, Y. (1985), 'Interpretation and misinterpretation of Jewish territorialism', in Newman (1985), pp. 111–24.

Shilhav, Y. (1993), 'Ethnicity and geography in Jewish perspective', *GeoJournal*, 30, pp. 273–7.

Shilhav, Y. (1995), 'Regionalism, territory and sovereignty: the Israeli scene', *Region and Regionalism*, 2, pp. 123–8.

Shilhav, Y. (1996), 'Judaism, geography, and sovereignty: the Middle East peace process as an educational dilemma', in Gradus and Lipshitz (1996), pp. 269–76.

Shilhav, Y. (1998), *Ultra-Orthodoxy in Urban Governance*, Jerusalem: The Floersheimer Institute for Policy Studies.

Shutte, A. (1993), *Philosophy for Africa*, Rondebosch: University of Cape Town Press.

Sibley, D. (1988), 'Purification of space', *Environment and Planing D: Society and Space*, 6, pp. 409–21.

Sibley, D. (1995), *Geographies of Exclusion: Society and Difference in the West*, London: Routledge.

Sikorski, R. (1997), *The Polish House: An Intimate History of Poland*, London: Weidenfeld & Nicolson.

Silberbauer, G. (1991), 'Ethics in small-scale societies', in P. Singer (1991), pp. 14–28.

Silk, J. (1998), 'Caring at a distance', *Ethics, Place and Environment*, 1, pp. 165–82.

Silk, J. (1999), 'The dynamics of community, place, and identity', guest editorial, *Environment and Planning A*, 31, pp. 5–17.

Simon, D. (1998), 'Rethinking (post)modernism, postcolonialism, and posttraditionalism: South–North perspectives', *Environment and Planning D: Society and Space*, 16, pp. 219–45.

Simon, D. and Dodds, K. (1998), 'Rethinking geographies of North–South development', *Third World Quarterly*, 19, pp. 595–606.

Singer, I. B. (1998), *Shadows on the Hudson*, New York: Farrar, Straus and Giroux.

Singer, P. (1972), 'Famine, affluence and morality', *Philosophy and Public Affairs*, 1, 229–43, reprinted in Laslett, P. and Fishkin, J. (eds), *Philosophy, Politics and Society*, 5th series, Oxford: Blackwell Publishers, pp. 21–35.

Singer, P. (ed.) (1991), *A Companion to Ethics*, Oxford: Blackwell Publishers.

Singer, P. (1993), review of Becker, L. C. and Becker, C. B. (1992), *Ethics*, 103, pp. 806–10.

Singer, P. (1995), *How are we to Live? Ethics in an Age of Self-Interest*, Amherst, NY: Prometheus Books.

Sitas, A. (1995), 'Exploiting Phumelele Nene: Postmodernism, intellectual work and ordinary lives', *Transformation*, 27, pp. 74–87.

Slater, D. (1997), 'Spatialities of power and postmodern ethics: rethinking geopolitical encounters', *Environment and Planning D: Society and Space*, 15, pp. 55–72.

Smith, D. M. (1965), *The Industrial Archaeology of the East Midlands*, Dawlish: David & Charles.

Smith, D. M. (1971), 'Radical geography: the next revolution?' *Area*, 3, 153–7.

Smith, D. M. (1973), *The Geography of Social Well-being in the United States: An Introduction to Territorial Social Indicators*, New York: McGraw-Hill.

Smith, D. M. (1977), *Human Geography: A Welfare Approach*, London: Edward Arnold.

Smith, D. M. (1979), *Where the Grass is Greener: Living in an Unequal World*, Harmondsworth: Penguin.

Smith, D. M. (1981), *Industrial Location: An Economic Geographical Analysis*, 2nd edition, New York: John Wiley.

Smith, D. M. (1985), *Apartheid in South Africa*, 3rd edition 1990, Cambridge: Cambridge University Press.

Smith, D. M. (1987), *Geography, Inequality and Society*, Cambridge: Cambridge University Press.

Smith, D. M. (1989), *Urban Inequality under Socialism: Case Studies from Eastern Europe and the Soviet Union*, Cambridge: Cambridge University Press.

Smith, D. M. (1994a), *Geography and Social Justice*, Oxford: Blackwell Publishers.

Smith, D. M. (1994b), 'Social justice and the post-socialist city', *Urban Geography*, 15, pp. 612–27.

Smith, D. M. (1994c), 'On professional responsibility to distant others', *Area*, 26, pp. 359–67.

Smith, D. M. (1995a), 'Geography, health and social justice: looking for the "right" theory', *Critical Public Health*, 6(3), pp. 5–11.

Smith, D. M. (1995b), 'Moral teaching in geography', *Journal of Geography in Higher Education*, 19, pp. 271–83.

Smith, D. M. (1997a), 'Back to the good life: towards an enlarged conception of social justice', *Environment and Planning D: Society and Space*, 15, pp. 19–35.

Smith, D. M. (1997b), 'Geography and ethics: a moral turn?' *Progress in Human Geography*, 21, pp. 583–90.

Smith, D. M. (1997c), 'Las dimensiones morales del desarrollo' [Moral dimensions of development], *Economica, Sociedad y Territorio*, 1, pp. 1–39.

Smith, D. M. (1998a), 'Geography and moral philosophy: some common ground', *Ethics, Place and Environment*, 1, pp. 7–34.

Smith, D. M. (1998b), 'How far should we care? On the spatial scope of beneficence', *Progress in Human Geography*, 22, pp. 15–38.

Smith, D. M. (1999a), 'Geography, community, and morality', *Environment and Planning A*, 31, pp. 19–35.

Smith, D. M. (1999b), 'Geography and ethics: how far should we go?' *Progress in Human Geography,* 23, pp. 116–22.

Smith, D. M. (1999c), 'Social justice and the ethics of development in post-apartheid South Africa', *Ethics, Place and Environment,* 2, pp. 157–77.

Smith, D. M. (2000a), 'Moral progress in human geography: transcending the place of good fortune', *Progress in Human Geography,* 24, pp. 1–18.

Smith, D. M. (2000b), 'Social justice revisited', *Environment and Planning A*, 32 (in press).

Smith, M. (1994), *The Moral Problem*, Oxford: Blackwell Publishers.

Smith, S. (1989), 'Society, space end citizenship: a human geography for the "new times"', *Transactions of the Institute of British Geographers*, 14, pp. 144–56.

Solomon, R. C. (1995), 'Justice as vengeance, vengeance as justice: a partial defence of Polymarchus', in Sterba et al. (1995), pp. 251–300.

Stein, S. M. and Harper, T. L. (1996), 'Planning theory for environmentally sustainable planning', *Geography Research Forum*, 16, pp. 80–101.

Steiner, G. (1997), *Errata: An Examined Life,* London: Weidenfeld & Nicolson.

Steinlauf, M. C. (1997), *Bondage to the Dead: Poland and the Memory of the Holocaust,* Syracuse: Syracuse University Press.

Sterba, J. P. (1981), 'The welfare rights of distant peoples and future generations: moral side-constraints on social policy', *Social Theory and Practice*, 7, pp. 99–119.

Sterba, J. P. (1996), 'Understanding evil: America slavery, the Holocaust, and the conquest of the American Indians', *Ethics*, 106, pp. 424–48 [review of Thomas, L. M. (1992), *Vessels of Evil: American Slavery and the Holocaust*, Philadelphia: Temple University Press].

Sterba, J. P. (1998), *Justice for Here and Now,* Cambridge: Cambridge University Press.

Sterba, J. P., Machan, T. R., Jagger, A. M., Galston, W. A., Gould, C. C., Fisk, M. and Solomon, R. C. (1995), *Morality and Social Justice: Point/Counterpoint*, London: Rowman & Littlefield.

Stern, P. (1991), 'Citizenship, community and pluralism: the current dispute on distributive justice', *Praxis International*, 13, pp. 261–97.

Sypnowich, C. (1993a), 'Justice, community, and antinomies of feminist theory', *Political Theory*, 21, pp. 484–506.

Sypnowich, C. (1993b), 'Some disquiet about "difference"', *Praxis International*, 13, 99–112.

Szewc, P. (1993), *Annihilation*, Norlam, IL: Dalkey Achive Press.

Szram, A. (1984), 'The impact of the investor on the work of the architect taking as examples the activities of factory owners in Łódź – I. K. Poznański and K. Scheibler', *Kwartalnik Architectury i Urbanistyki*, 29, 1–2, pp. 281–7.

Tarkhan-Mouravi, G. (1998), 'Some aspects of poverty in Georgia', in Melikidze, V. and Smith, D. M. (eds), *Spatial Problems of the Transition to a Market Economy and Democratisation: Theory and Practice*, Proceedings of the Third British-Georgian Geographical Seminar, London: Queen Mary and Westfield College, pp. 89–104.

Taylor, P. J. (1994), 'The state as container: territoriality in the modern world-system', *Progress in Human Geography*, 18, pp. 151–62.

Taylor, P. J. (1995), 'Beyond containers: internationality, interstateness, interterritoriality', *Progress in Human Geography*, 19, pp. 1–15.

Taylor, P. J. (1996), 'Territorial absolutism and its evasions', *Geography Research Forum*, 16, pp. 1–12.

Tec, N. (1986), *When Light Pierced the Darkness: Christian Rescue of Jews in Nazi-Occupied Poland*, Oxford: Oxford University Press.

Tec, N. (1989), 'Of help, understanding, and hope: righteous rescuers and Polish Jews', *Polin: A Journal of Polish-Jewish Studies*, 4, pp. 296–310.

The Editors (1991), 'Editorial', *Theoria*, 78, pp. i–ii.

The Jerusalem Post, Internet Edition, www.jpost.com

Thompson, J. (1992), *Justice and World Order: A Philosophical Inquiry*, London: Routledge.

Todorov, T. (1999), *Facing the Extreme: Moral Life in the Concentration Camps*, London: Weidenfeld & Nicolson.

Tomaszewski, J. (1991), 'Jews in Łódź in 1931 according to statistics', *Polin: A Journal of Polish-Jewish Studies*, 6, pp. 173–200.

Trachtenberg, Z. (1997), 'The takings clause and the meanings of land', in Light and Smith (1997), pp. 63–90.

Tronto, J. (1987), 'Beyond gender difference to a theory of care', *Signs: Journal of Women in Culture and Society*, 12, pp. 645–63.

Tronto, J. (1993), *Moral Boundaries: A Political Argument for an Ethic of Care*, London: Routledge.

Tuan, Y.-F. (1986), *The Good Life*, Madison, WI: The University of Wisconsin Press.

Tuan, Y.-F. (1989), *Morality and Imagination: Paradoxes of Progress*, Madison, WI: The University of Wisconsin Press.

Tuan, Y.-F. (1993), *Passing Strange and Wonderful: Aesthetics, Nature and Culture*, Washington DC: Island Press.

Tucker, A. (1994), 'In search of home', *Journal of Applied Philosophy*, 11, pp. 181–7.

Turner, B. S. (1986), *Equality*, Chichester: Ellis Horwood.

Unger, M. (ed.) (1995), *The Last Ghetto: Life in the Łódź Ghetto 1940–1944*, Jerusalem: Yad Vashem.

Valentine, G. (1993), '(Hetero)sexing space: lesbian perceptions and experiences of everyday spaces', *Environment and Planning D: Society and Space*, 9, pp. 395–413.

Valentine, G. (1996), 'Angels and devils: moral landscapes of childhood', *Environment and Planning D: Society and Space*, 14, pp. 581–99.

van der Berg, S. (1991), 'Redirecting government expenditure', in Moll, P., Natrass, N. and Loots, L. (eds), *Redistribution: How Can it Work in South Africa?* Cape Town: David Philip, pp. 74–85.

Vetlesen, A. J. (1993), 'Why does proximity make a moral difference? Coming to terms with lessons learned from the Holocaust', *Praxis International*, 12, pp. 371–86.

Waldron, J. (1992), 'Superseding historic injustice', *Ethics*, 103, pp. 4–28.

Walker, G. (1998), 'Environmental justice and the politics of risk', *Town and Country Planning*, December, pp. 358–9.

Walzer, M. (1983), *Spheres of Justice: A Defence of Pluralism and Equality*, Oxford: Basil Blackwell.

Walzer, M. (1990), 'The communitarian critique of liberalism', *Political Theory*, 18, pp. 6–23.

Walzer, M. (1994), *Thick and Thin: Moral Argument at Home and Abroad*, Notre Dame and London: University of Notre Dame Press.

Warburton, D. (1998), *Community and Sustainable Development: Participation in the Future*, London: Earthscan.

Waterman, S. (1999), 'Keeping a distance: Israel at 50', *Political Geography*, 18, pp. 115–29.

Whatmore, S. (1997), 'Dissecting the autonomous self: hybrid cartographies for a relational ethics', *Environment and Planning D: Society and Space*, 15, pp. 37–53.

Williams, B. (1972), *Morality: An Introduction to Ethics*, Cambridge: Cambridge University Press.

Williams, B. (1985), *Ethics and the Limits of Philosophy*, London: Fontana Press/Collins.

Wilson, R. A. (1996), 'The siswe will not go away: the Truth and Reconciliation Commission, human rights and nation-building in South Africa', *African Studies*, 55(2), pp. 1–20.

Wolaniuk, A. (1997), 'Spatial distribution of crime in Łódź and its urban region', in Liszewski and Young (eds) (1997), pp. 203–13.

Wolpe, H. (1995), 'The uneven transformation from apartheid in South Africa', *Transformation*, 27, pp. 88–101.

Wong, D. (1993), 'Relativism', in P. Singer (1991), pp. 442–50.

Young, I. M. (1990a), 'The ideal of community and the politics of difference', in Nicholson (1990), pp. 300–23.

Young, I. M. (1990b), *Justice and the Politics of Difference*, Princeton, NJ: Princeton University Press.

Young, I. M. (1997), 'Unruly categories: a critique of Nancy Fraser's dual systems theory', *New Left Review*, 222, pp. 147–60.

Young, J. E. (1993), *The Texture of Memory: Holocaust Memorials and Meanings*, New Haven and London: Yale University Press.

Young, M. and Willmott, P. (1957), *Family and Kinship in East London*, London: Routledge & Kegan Paul.

INDEX

Page references in *italics* indicate figures.